RUSSIAN NAT

Russian Nature
Exploring the Environmental Consequences of Societal Change

JONATHAN D. OLDFIELD
University of Birmingham, UK

Routledge
Taylor & Francis Group

LONDON AND NEW YORK

First published 2005 by Ashgate Publishing

2 Park Square, Milton Park, Abingdon, Oxon OX14 4RN
711 Third Avenue, New York, NY 10017, USA

Routledge is an imprint of the Taylor & Francis Group, an informa business

First issued in paperback 2016

British Library Cataloguing in Publication Data
Oldfield, Jonathan D.
    Russian nature : exploring the environmental consequences
    of societal change. - (Ashgate studies in environmental
    policy and practice)
    1. Environmental protection - Social aspects - Russia
    (Federation) 2. Environmental policy - Russia (Federation)
    3. Russia (Federation) - Environmental conditions
    I. Title
    333.7'2'0947

Library of Congress Cataloging-in-Publication Data
Oldfield, Jonathan D.
    Russian nature : exploring the environmental consequences of societal change / by Jonathan D. Oldfield.
        p. cm. -- (Ashgate studies in environmental policy and practice)
    Includes bibliographical references and index.
    ISBN 0-7546-3940-1
    1. Russia (Federation)--Environmental conditions. 2. Environmental policy--Russia (Federation) 3. Social change--Russia (Federation) I. Title. II. Series.

    GE160.R8O53 2006
    363.7'00947--dc22

                                                    2005014422

ISBN 978-0-7546-3940-4 (hbk)
ISBN 978-1-138-27800-4 (pbk)

Transferred to Digital Printing in 2014

Contents

Map of the Russian Federation

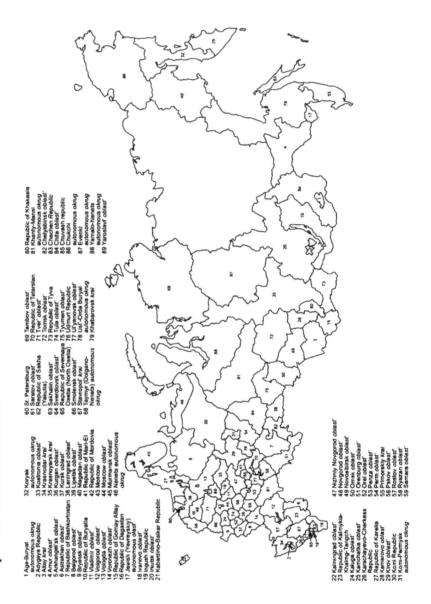

1 Aga-Buryat autonomous okrug
2 Adygeya Republic
3 Altay krai
4 Amur oblast'
5 Arkhangel'sk oblast'
6 Astrakhan oblast'
7 Republic of Bashkortostan
8 Belgorod oblast'
9 Bryansk oblast'
10 Republic of Buryatia
11 Vladimir oblast'
12 Volgograd oblast'
13 Vologda oblast'
14 Voronezh oblast'
15 Republic of Gornay Altay
16 Republic of Dagestan
17 Jewish (Yevreyskiy) autonomous oblast'
18 Ivanovo oblast'
19 Ingush Republic
20 Irkutsk oblast'
21 Kabardino-Balkar Republic

22 Kaliningrad oblast'
23 Republic of Kalmykia-Khalmg-Tangch
24 Kaluga oblast'
25 Kamchatka oblast'
26 Karachayevo-Cherkess Republic
27 Republic of Karelia
28 Kemerovo oblast'
29 Kirov oblast'
30 Komi Republic
31 Komi-Permyak autonomous okrug

32 Koryak autonomous okrug
33 Kostroma Republic
34 Krasnodar krai
35 Krasnoyarsk krai
36 Kurgan oblast'
37 Kursk oblast'
38 Leningrad oblast'
39 Lipetsk oblast'
40 Magadan oblast'
41 Republic of Mari-El
42 Republic of Mordovia
43 Moscow
44 Moscow oblast'
45 Murmansk oblast'
46 Nenets autonomous okrug

47 Nizhniy Novgorod oblast'
48 Novgorod oblast'
49 Novosibirsk oblast'
50 Omsk oblast'
51 Orenburg oblast'
52 Orel oblast'
53 Penza oblast'
54 Perm oblast'
55 Primorskiy krai
56 Pskov oblast'
57 Rostov oblast'
58 Ryazan oblast'
59 Samara oblast'

60 St. Petersburg
61 Saratov oblast'
62 Republic of Sakha (Yakutia)
63 Sakhalin oblast'
64 Sverdlovsk oblast'
65 Republic of Severnaya Osetia (North Osetia)
66 Smolensk oblast'
67 Stavropol' krai
68 Taymyr (Dolgano-Nenets) autonomous okrug

69 Tambov oblast'
70 Republic of Tatarstan
71 Tver' oblast'
72 Tomsk oblast'
73 Republic of Tyva
74 Tula oblast'
75 Tyumen oblast'
76 Udmurt Republic
77 Ul'yanovsk oblast'
78 Ust'-Orda Buryat autonomous okrug
79 Khabarovsk krai

80 Republic of Khakasia
81 Khanty-Mansi autonomous okrug
82 Chelyabinsk oblast'
83 Chechen Republic
84 Chita Republic
85 Chuvash republic
86 Chukchi autonomous okrug
87 Evenki autonomous okrug
88 Yamalo-Nenets autonomous okrug
89 Yaroslavl' oblast'

List of Figures

List of Tables

List of Plates

Acknowledgements

The origins of this study lie in doctoral work carried out during the late 1990s. This was funded by a scholarship from the UK Economic and Social Research Council (ESRC), which provided me with the opportunity to spend extended periods carrying out fieldwork in the Russian Federation. I am indebted to the ESRC for its initial willingness to fund the research as well as its continued support of related research initiatives.

I would like to extend my thanks to Denis Shaw who was instrumental in cultivating my initial interest, respect and appreciation of Russian history and culture. Furthermore, his constant readiness to reflect upon the complexities of contemporary Russian society has provided the basis for many interesting and thought-provoking discussions over the years. Mike Bradshaw also offered guidance and support during the course of my doctorate for which I am very grateful. My time spent working in the Baykov library at the Centre for Russian and East European Studies (University of Birmingham) is greatly missed. Nigel Hardware, Graham Dix, Mike Berry, Marea Arries and Tricia Carr were unfailingly helpful and sympathetic to my many requests. I was fortunate in that my association with CREES coincided with an exceptional group of graduate students. Rod Thornton, Kate Thompson, Erica Richardson, John Round, Luke March, Alison Stenning and Moya Flynn gave invaluable support during these years and were an unwavering source of inspiration and motivation. Luckily, I have been able to continue working with a number of these individuals in recent years. In particular, the careful and perceptive criticism of Moya Flynn has furthered considerably my own insight and understanding of change within contemporary Russian society.

The lively intellectual environments characterising the School of Geography at Nottingham University and the Department of Geography at Queen Mary, University of London, encouraged me to place my work within wider contexts and, at the same time, question many of my underlying assumptions. I am especially grateful to Adam Fagan and Adam Swain for their enthusiasm to discuss ideas related to various aspects of post-socialist change, and for introducing me to new literatures. More generally, the support of Alastair Owens, Sarah O'Hara and Miles Ogborn is very much appreciated.

During fieldwork periods in Russia, I met many individuals who gave freely of their time and I am grateful to them all for their help, critical insight, patience and, principally, their friendship. Special mention needs to be made of the following: Sergei Gorshkov, Elena Surovikina, Tatiana Nefedova, Andrei Treivish, Oksana Boitsova, Yulia Odnakova and Zhene Smirnovoi. In addition, Oleg

Yanitsky and Anna Kuz'mina have been a constant source of support and guidance during the last five years, and their eagerness to discuss diverse themes and topics has been invaluable in advancing my own understanding of issues and events.

A range of people have provided constructive criticism, or else much needed encouragement, at various times during the course of the last few years. In no particular order, I would like to thank: Nat Trumbull, Petr Pavlinek, Jessica Graybill, D.J. Peterson, Tassilo Herrschel as well as others associated with the Russian, Central Eurasian, and Eastern European Specialty Group (RCEEE) of the Association of American Geographers and the Post-Socialist Geographies Research Group (PSGRG) of the Royal Geographical Society. Furthermore, Nobuko Ichikawa, Emma Wilson and Andy Tickle have been more helpful than they can possibly realise.

The process of putting together the manuscript was facilitated greatly by the technical assistance of Kevin Burkhill and Lydia Buravova and I thank them both for their high level of professionalism and friendship. Furthermore, Val Rose at Ashgate Press was an enormous help during those periods when work on the book became more difficult. Parts of this study draw from previously published work and I am grateful to John Wiley and Sons Ltd. (*European Environment*) and Taylor and Francis (*Environmental Politics* and *Post-Communist Economies*) for permission to reuse material published in their journals.

Finally, I would like to thank my family for their constant support and understanding, which has eased considerably the sometimes onerous task of bringing this work together. While acknowledging the influence and assistance of those mentioned above, I am, of course, solely responsible for any mistakes, errors and misinterpretations found within the study.

Jonathan Oldfield
London

The author and the publisher gratefully acknowledge permission for use of the following material:

Oldfield, J.D. (2002), 'Russian environmentalism', *European Environment*, John Wiley and Sons Ltd, volume 12, pp. 117-129.

Oldfield, J.D. (2001), 'Russia, systemic transformation and the concept of sustainable development', *Environmental Politics*, Taylor and Francis, volume 10: 3, pp. 94-110. <http://www.tandf.co.uk>

Oldfield, J.D. (2000), 'Structural economic change and the natural environment in the Russian Federation', *Post-Communist Economies*, Taylor and Francis, volume 12: 1. <http://www.tandf.co.uk>

Glossary

CEE	Central and eastern Europe
EBRD	European Bank for Reconstruction and Development
EEA	European Environment Agency
EU	European Union
FSU	Former Soviet Union
GEF	Global Environment Facility
GDP	Gross domestic product
Gosgidromet SSSR	State Hydrometerological Service SSSR
Gosgortekhnadzor	State Mining and Industrial Inspectorate
Goskomekologiya	State Committee for Environmental Protection
Goskompriroda SSSR	State Committee for the Protection of Nature SSSR
Goskokmstat	State Statistical Committee
Gossanepidsluzhba	Federal Centre for Sanitary Epidemiological Inspection
IZA	Air pollution index (*Indeks zagryazneniya atmosfery*)
Minzdrav	Ministry for Health
MPC	Maximum permissible concentration
MPR	Ministry for Natural Resources (*Ministerstvo prirodnykh resursov*)
MSAP	Mobile source air pollution
OECD	Organisation for Economic Co-operation and Development
PDD	Polluted drainage discharge
Rosgidromet	Federal Service for Hydrometeorology and Environmental Monitoring
RSFSR	Russian Soviet Federative Socialist Republic
SSAP	Stationary source air pollution
SSSR	Union of Soviet Socialist Republics
TACIS	Technical Assistance to the Commonwealth of Independent States (EU technical assistance programme)
TETs	Thermal power station (*teploelektrotsentral'*)
UNCED	United Nations Conference for Environment and Development
UNDP	United Nations Development Programme
UNECE	United Nations Economic Commission for Europe
UNEP	United Nations Environment Programme

UNFCCC	United Nations Framework Convention on Climate Change
WHO	World Health Organisation
WSSD	World Summit for Sustainable Development

A Note Concerning Transliteration

The transliteration system used in this thesis is based on a modified Library of Congress system utilised by the Centre for Russian and East European Studies, University of Birmingham, UK. The English spelling of words such as common place names has been retained.

Cyrillic symbol	Transliteration
а	a
б	b
в	v
г	g
д	d
е	e
ж	zh
з	z
и	i
й	i
к	k
л	l
м	m
н	n
о	o
п	p
р	r
с	s
т	t
у	u
ф	f
х	kh
ц	ts
ч	ch
ш	sh
щ	shch
ъ	"
ы	y
ь	'
э	e
ю	yu
я	ya

Chapter 1

Introduction

Introduction

Towards the end of April 1986, an event took place on the western borders of the former Soviet Union which was to have a profound impact on the lives of millions of Soviet citizens, and touched the lives of countless more across the European continent and beyond. The explosion of the number four reactor at the Chernobyl' nuclear power station, and the subsequent mishandling of the situation by the Soviet authorities, was considered by many at the time as symbolic of the malaise at the heart of the Soviet Union's socialist project. For Soviet citizens, it provided evidence of a system in decline and governed by a political machine incapable of protecting their own interests. For the Western public, the sight of the devastated reactor accompanied by satellite-assisted imagery of radioactive plumes spreading across the European continent, reinforced notions of deep social, economic, political and environmental flaws extant within the Soviet system.

The conceptualisation of the Soviet development model, as both economically inefficient and highly polluting, emerged powerfully in Western consciousness during the late 1960s and early 1970s with the publication of groundbreaking studies by such authors as Goldman (1972), Pryde (1972) and the dissident Zeev Wolfson (publishing under the pseudonym Boris Komarov, 1978/1980). It was further supported by a number of works published towards the end of the Soviet period (e.g. Massey Stewart, 1992; Peterson, 1993; Pryde, 1991; see also Singleton, 1976). These negative perceptions were substantiated during the 1980s by environmental disasters, such as the aforementioned Chernobyl' accident and the marked shrinkage of the Aral Sea, in combination with the particularities of the Soviet Union's decline whereby environmental grievances formed a focal point for public protest across the region. More broadly, the late 1980s was also a period of heightened environmental awareness within the West, coinciding with publication of the Brundtland report '*Our Common Future*' (WCED, 1987) and preparations for the 1992 United Nations Conference on Environment and Development (UNCED). These events in concert helped to reinforce a rather undifferentiated vision of the Soviet Union's environmental situation, which belied underlying variation in both the nature and the extent of environmental issues at the sub-national scale. Notions of widespread environmental crisis spilled over somewhat inevitably into the post-Soviet period and were easily assimilated by the neoliberal rhetoric dominant during the early 1990s. This is not to cloud the fact that the Soviet system bequeathed extensive environmental problems with global ramifications, and that these remain relevant in the current period. Nevertheless, there is a need to move purposefully beyond broad understandings of the Soviet environmental legacy in order to engage with the particularities of the Russian Federation's contemporary environmental situation. The last decade has witnessed a marked reworking of the relationship between Russian society and the wider environment, and yet much of our understanding remains generalised in its nature and characterised by limited engagement with, and appreciation of, underlying regional variation. Furthermore, while issues such as nuclear waste containment and storage, which have their roots embedded in the logic of the Soviet system, continue to receive significant levels of popular attention within the West, the regional and global consequences of Russia's rising motor vehicle numbers, the loss of biodiversity in the Russian Far East region or the widespread illegal exploitation of natural resources, are given rather less consideration. This reflects, in part, the strategic environmental agendas of Western countries and coalitions, since they play a leading role in determining the international agenda. For example, the European Union's (EU) sensitivity towards Russia's nuclear waste (both domestic and military) is closely associated with the perceived threat it poses for bordering EU countries (e.g. European Commission, 2001a).

In recognition of the prevailing generalisations, this study explores key elements of Russia's contemporary environmental situation, at both the national and regional level, through an engagement with aspects of the country's dynamic social, economic and political circumstances. This introductory chapter provides a contextual basis for the study and is divided into four main parts. First, the different ways in which the West has approached Russia's environmental situation since 1991 are explored. More specifically, it is suggested that at least three dominant conceptual frameworks are recognisable related to notions of environmental crisis and environmental improvement in addition to Russia's role as a global environmental player. Second, and in recognition of the intimate connections between prevailing societal trends and resulting environmental pressures, the chapter examines current understandings related to the transformation of post-socialist societies and the implications that these have for our interpretation of associated environmental trends. Third, and in order to draw attention to the wider context in which Russia's societal changes are taking place, the environmental implications of Russia's progressive entrenchment in the flows and processes of globalised economic systems are considered. Fourth, the substantive aims of the study are outlined and underlying methodological considerations discussed.

Conceptualising Environmental Issues Within the Former Soviet Union

As indicated above, Western academic interest in the environmental problems of the former Soviet Union dates back at least 40 years. This interest has been generally critical of the environmental consequences of the Soviet development model. A powerful image of environmental degradation has emerged articulated through a number of case studies, in addition to those already mentioned, such as soil erosion in southern European Russia or the pollution threat to rivers and large water resources (e.g. the river Volga, the Black Sea and Lake Baikal). Furthermore, this general perception is supported by a coherent body of causal factors ranging from critiques of Marxist-Leninist ideals and, in particular, their interpretation by Stalin, through to the nature of the prevailing Soviet administrative and bureaucratic system (see Chapter 2). The manner in which Western academics comprehended the Soviet environment was influenced to some extent by the Soviet regime's limited explicit engagement with environmental issues. For example, while Western societies were experiencing an environmental awakening during the 1970s, the Soviet regime remained guarded about the state of its air, water and land resources, both domestically and internationally. Furthermore, it tended to adopt a cautious approach with regard to international environmental agreements, although strategic environmental cooperation initiatives were a feature of the 1970s and 1980s as Brezhnev, and later Gorbachev, searched for symbolic displays of collaboration with the West (Darst, 2001, pp. 21-28). The English language publication of the book '*The destruction of nature in the Soviet Union*' in 1980 by Zeev Wolfson served to heighten the notion that the Soviet Union's environmental problems were far more serious than hitherto imagined. Importantly, the secrecy surrounding the state of the Soviet environment only reinforced the belief that the Soviet Union had something to hide whereas the roots of this concealment were arguably associated with more general geopolitical and ideological concerns. The West's preoccupation with particular elements of the Soviet Union's environmental problems, combined with the difficulty of breaching the East-West divide, tended to preclude a detailed engagement with Soviet scholarly debate related to environmental protection issues. An exception to this was the highly illuminating study by Joan DeBardeleben (1985), which endeavoured to elucidate the complexity of academic deliberation within both the Soviet Union and the GDR, and indicated the need for a more sophisticated approach towards the environmental failings of the Soviet development model (see Chapter 2). Other studies worthy of mention at this juncture include the work of geographers such as Matley (1966; 1982) and Hooson (1968) and their engagement with the ideas of Soviet geographers, which provide useful insights into debates concerning the interaction between society and nature during the Soviet period.

Since the fall of the Soviet Union, much of the literature concerned with environmental issues has tended to remain preoccupied with the strained nature of the regional ecological situation, a significant element of which has its origins in the excesses of the Soviet development model (e.g. Bridges and Bridges, 1996; Feshbach and Friendly, 1992; Peterson, 1993; Saiko, 2001). While these studies have been important in highlighting the range and extent of such problems, there is a need to move on from conceptualisations of environmental disaster, embedded within the logic of the Soviet system, in order to explore more fully emerging contemporary trends. Josh Newell's book on the Russian Far East (Newell, 2004) is indicative of such an approach with its exploration of emerging conservation and development issues. Furthermore, Newell's deliberate focus on a discrete region of Russia is useful in drawing attention to the country's considerable size and regional heterogeneity. Recent historical and archival work has also started to add new layers of understanding, revealing the complexity of Soviet approaches to environmental issues embedded deep in Soviet society and with linkages to the pre-revolutionary period. In particular, the work of Weiner (1988; 1999) has helped to undermine the generalised view of the Soviet Union as an environmental 'basket-case' by emphasising the continuity of environmental thought between pre-revolutionary Russia and the Soviet period. In a similar vein, authors such as Yanitsky (1993; 2002) and Mirovitskaya (1998) have exposed the historical roots of Russia's environmental movement (see Chapters 2 and 5). Yanitsky's sociological approach (e.g. Yanitsky, 1993) is especially useful in stressing the importance of the Soviet experience in shaping contemporary approaches to environmental issues amongst prominent Russian environmentalists. The emergence of such work ensures that the tendency towards overly crude interpretations of Soviet environmental degradation is checked and, rather than dismiss Soviet society as devoid of meaningful conservation and environmental activity, we are encouraged to identify reasons why environmental sensibilities apparent during this period could not flourish, and to pursue further the nature of such sensibilities. This interest in the continuity and development of environmental thought dovetails with the aforementioned work of Hooson (1968) and his exploration of the evolution of Russian geographical ideas from the 18th Century through to the mid-Soviet period (see also Oldfield and Shaw, 2002).

Western 'Imaginings' of the Russian Environment

During the course of the 1990s, at least three, often competing, approaches to the Russian environment emerged and these remain influential to varying degrees at the present time. Drawing from the discussion in the previous section, the first of these has its roots in the Soviet period and is grounded on notions of environmental crisis and disaster associated with an extensive military-industrial complex and associated technogenic failings. A second approach to the Russian environment is related closely to the powerful transition discourse that took shape in the years following the break-up of the Soviet Union (see below). This focuses attention on the anticipated short- to medium-term improvements in the overall efficiency of the Russian economy allied to the progressive implantation of capitalistic relations and associated falls in the country's overall pollution load. Within such a conceptual framework, improving environmental trends emerged as a means for assessing the 'success' of economic transition and the experiences of other post-socialist countries, such as Poland and Hungary, were posited as supportive of such an approach (e.g. Zamparutti and Gillespie, 2000). In a limited sense, this framework is useful in drawing attention to the relationship between ongoing macro-economic restructuring and certain environmental trends. For example, there is considerable scope for improving the efficiency of the Russian production system with concomitant benefits for rates of pollution output and resource-use. However, a main problem with this type of explicatory framework relates to its blinkered engagement with processes of societal transformation evident within the Russian Federation, and the ease with which it denies alternative futures for Russia. Thus, the employment of models of transition, based on idealised notions of how a market economy works, forecloses critical debate concerning the environmental consequences of the restructuring process at a variety of scales. Any failing is comprehended as evidence of the inadequate implementation of market infrastructure rather than a sign of the capitalist system's underlying weaknesses. By extension, this type of approach provides limited room for engaging with larger questions concerning the viability of the capitalist model over the long-term (e.g. see Beck, 1996; Foster, 2002; Porritt, 1984). Additionally, there is the danger that environmental change is reduced to a series of responses stimulated by prevailing social, economic and political tendencies, thus ignoring the highly complex (non-linear) and often unpredictable nature of biophysical systems.

Some of the conceptual deficiencies outlined above are addressed by a third identifiable approach to the Russian environment evident during the 1990s and related to Russia's significance as an environmental actor at the global scale (e.g. Darst, 2001; Hønneland, 2003; Oldfield et al., 2003). Such an approach begins to move beyond the restricted notions of economic optimisation and environmental improvement and emphasises Russia's role for the enforcement of international environmental agreements as well as the effective functioning of global ecological systems. More importantly, such a discourse distances itself, at least in part, from the assuredness of the transition model of change and draws attention to the uncertainties surrounding Russia's, and indeed the Earth's, environmental state over the medium- to long-term. For example, Russia's prominence with regard to global greenhouse gas emissions has been the subject of intense scrutiny in recent years due to its initial reluctance to ratify the Kyoto agreement on climate change (Oldfield et al., 2003). This approach to Russia's contemporary environmental situation has strong connections with the general literature concerned with environmental risk and global environmental change, and is shaped by prevailing understandings of what constitutes a global environmental issue (e.g. climate change, biodiversity loss etc.). These understandings are determined to a large extent by the complex interplay of political, scientific and non-governmental actors at the supranational scale. Nevertheless, Russia's considerable size (covering more than 11 percent of the Earth's land surface), coupled with its relatively extensive industrial base, ensures that it plays a significant role with respect to most, if not all, environmental criteria regularly cited in generic State of the Environment documentation. The recent Global Environment Outlook 3 report (UNEP, 2002), brought together under the auspices of the United Nations Environment Programme (UNEP), is indicative of the generally accepted thematic approach to environmental concerns for international comparison with its coverage of land, air, water, biodiversity and forest issues. Furthermore, the Russian Federation features prominently in all sections. Similarly, the Environmental Programme for Europe, driven by the United Nations Economic Commission for Europe (UNECE), has presided over the production of a number of comprehensive environmental surveys of the European continent since 1991, and these have included countries of the former Soviet Union. Such surveys highlight Russia's continental significance with respect to issues such as hazardous waste production, Arctic Sea water pollution and levels of biodiversity loss (EEA, 2003).

In recent years more sophisticated approaches to understanding environmental issues within the post-socialist countries of central and eastern Europe (CEE) and the former Soviet Union (FSU) have been advanced, largely in response to a recognised under-theorising in this area (e.g. Herrschel and Forsyth, 2001; Pavlinek and Pickles, 2000; Tickle, 2000). To a large extent, such approaches have been grafted on to existing conceptual frameworks characterised by a desire to understand societal change within the region.[1] The deliberate engagement with advances in social science thinking in order to further understanding of the region's changing environmental situation has considerable merit due to the close relationship between societal change and key processes operating within the wider environment. Nevertheless, this can still preclude an engagement with the aforementioned non-linear responses of biophysical systems to changing societal configurations (see Forsyth, 2003; Harvey, 1996; Pavlinek and Pickles, 2000). Thus, it can be argued that a truly comprehensive approach to the environmental implications of Russia's societal changes demands extensive analysis of *both* social trends *and* natural biophysical systems at a variety of scales. While such an approach is beyond the scope of this study, the contemporary environmental situation is arguably made more understandable through an engagement with socio-economic trends and changing structures of governance. In recognition of this, the following section explores the main debates within current social science literature concerned with the reworking of post-socialist societies, and associated implications for our understanding of environmental change within the region.

Societal Change and the Environment in CEE and the FSU

The social science literature concerned with post-socialist change is highly diverse, and yet it is possible to make certain generalisations. Indeed, there has been a tendency in recent years to perceive the literature as divided into two distinct parts characterised by fundamentally different ontological frameworks and approaches to the processes of change within CEE and the FSU (see Bradshaw and Stenning, 2004, pp. 12-14). In reality, the many different interpretations of post-socialist change represent a far more complex and convoluted theoretical landscape. Nevertheless, while such a high level of generalisation should be employed with caution, it can provide a starting point from which to engage with the extensive literature concerning societal transformation within the former socialist countries of CEE and the FSU.

An influential framework for understanding change within post-socialist societies prominent during the early 1990s, and that continues to underpin many contemporary assessments of the region, is characterised by what might be termed a temporal understanding of space. In this sense, past, present and future societal trends within the region are rendered knowable in relation to the experiences of

---

[1] See the edited volume by Pickles and Smith (1998) for a critical approach to the nature of post-socialist change.

advanced Western economies and thus imbued with Western mores and values (see Massey, 2005, p. 68). Furthermore, there is a tendency to believe in, or else tacitly accept, 'capitalism by design', which contends that effective market-type economies can be constructed according to the imposition of particular arrangements of institutional frameworks and management systems. Taken to its extreme, this approach can, unwittingly or not, treat the post-socialist space as a 'clean slate' or *tabula rasa* on which new futures can be constructed unhindered by the past (Bradshaw and Stenning, 2004, pp. 13-14; Swain, 2002). While many commentators resist such conceptualisations of space, it is interesting to note how easily working assumptions are often reducible to such crudities. For example, Pickvance (2003, p. 9) suggests that the notion of 'transformation', which has been employed as a less restrictive and deterministic term than 'transition' when describing ongoing change within CEE and the FSU, still '...implies extensive change and also prejudges the character of change'. Linear understandings of societal change, associated closely with Western experience, have formed the basis for working through notions of concomitant environmental improvement within former socialist countries (see diagram A, Figure 1.1). Such approaches find support, in part at least, within certain areas of contemporary environmental thought. For example, the concept of ecological modernisation has received critical attention during the course of the last decade with its challenge to the commonly held position that economic growth and environmental improvement are mutually exclusive development aims (see Murphy, 2000). Sections of this literature explore the various ways in which institutional, technical and policy innovations can encourage improvements in environmental performance within the context of specific industrial sectors or national economies.[2] In many instances, the required innovations are based heavily on the experience of Western societies (and, indeed, largely the experience of European countries) thus raising legitimate concerns over their relevance for other parts of the globe. The underlying impetus of ecological modernisation has overlap with some of the core tenets of sustainable development, where an emphasis is placed on the 'greening' of industrial production systems and economic processes in tandem with social and environmental betterment (see Chapter 4). Indeed, sustainable development has emerged as a key concept underpinning the activities of main international actors, lending institutions and other inter-governmental actors working within CEE and FSU. For example, the European Bank for Reconstruction and Development (EBRD) incorporated the notion of sustainable development into its founding mandate, and the World Bank recently revamped its operations in this regard with the establishment of the Environmentally and Socially Sustainable Development (ESSD) Vice Presidency.[3] It can be argued that a major weakness of such

---

[2] See the collection of articles in the journal *Geoforum* (Vol. 31, No. 1, 2000) for an overview of critical work in this area.

[3] See the World Bank publication: Sustainable Development (ESSD) Vice Presidency: Reference guide. Available at:
http://lnweb18.worldbank.org/ESSD/sdvext.nsf/43ByDocName/WorldBankSustainableDev elopmentReferenceGuide (accessed December 2004).

operational and policy commitments lies in their largely uncritical acceptance of the Western development model as a basis for achieving long-term reconciliation between societal development and effectively functioning ecological systems. As such, they discourage an engagement with wider discourses of regional and global environmental change and associated critical literatures, which cast doubt on the environmental efficacy of the capitalist model over the long-term. Furthermore, and as mentioned above, their inherently limited understanding of societal change downplays the likelihood that countries such as Russia might follow alternative societal paths associated with specific environmental consequences.

Large swathes of the critical social science literature have resisted the limited explicatory framework outlined above in order to posit a more nuanced and regionally differentiated understanding of the post-socialist region. At a general level, this emphasises the composite nature of contemporary society as well as the openness of future development trajectories at a variety of scales and has a range of implications for our understanding of related environmental trends (see Figure 1.1, Diagram B). Furthermore, this literature has been extremely useful in exposing the underlying tensions evident in prioritising trends of disjuncture over continuity. In other words, there has been a concerted effort to move beyond the simple binary of 'socialist' and 'post-socialist' in order to explore the particularities of socialist and contemporary social phenomena as well as the connections and linkages between the two periods. This is evident in the case of ethnographic type studies where the focus on the 'every day' ensures that continuities with past practices are more easily prioritised (e.g. Burawoy and Verdery, 1999; Flynn, 2004; Hann, 2002). In addition, such linkages are also apparent at the institutional level with contemporary modes of behaviour mired, to varying degrees, in socialist practices (e.g. Clark and Soulsby, 1999; Swain, 2002). An understanding of the way in which elements of the past combine and merge with features of the contemporary period has been the focus of much work and speculation during recent years (e.g. Grabher and Stark, 1997; Smith and Swain, 1998, pp. 26-27; Stark and Bruszt, 1998). With respect to the environment, Pavlinek and Pickles (2000) are particularly effective in highlighting the ability of socialist environmental practices to remain influential in CEE post-1989. At the same time, other commentators underline the need for more empirical work in order to detect the precise ways in which elements of the past and present interact within post-socialist societies (e.g. Pickvance, 2003, p. 10).

As indicated above, attempts have been made to think through some of the general themes extant within the critical social science literature in relation to the environmental situation of post-socialist countries. For example, the aforementioned work of Pavlinek and Pickles (2000) provides an interesting set of analytical and conceptual tools with which to start to 'unpick' prevailing stereotypes. Their exploration of uneven development is particularly effective in helping to expose the regionally varied nature of environmental change within former socialist countries of CEE related to the imposition of capitalistic relations. Tickle (2000) augments the work of Pavlinek and Pickles with his application of regulation theory to nature conservation in the Czech Republic, and by drawing attention to how socialist practices have been maintained to varying degrees in the

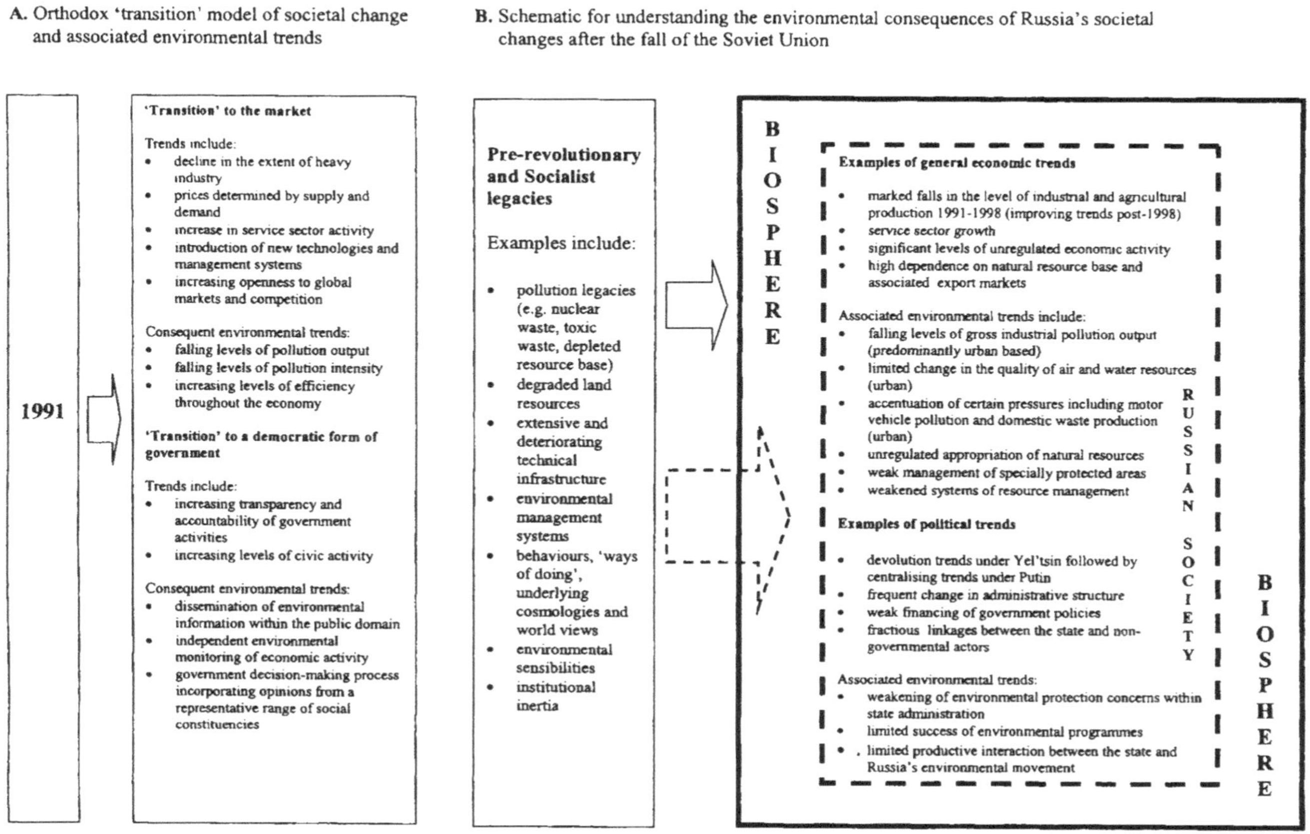
A. Orthodox 'transition' model of societal change and associated environmental trends
B. Schematic for understanding the environmental consequences of Russia's societal changes after the fall of the Soviet Union
1991
'Transition' to the market
Trends include:
• decline in the extent of heavy industry
• prices determined by supply and demand
• increase in service sector activity
• introduction of new technologies and management systems
• increasing openness to global markets and competition
Consequent environmental trends:
• falling levels of pollution output
• falling levels of pollution intensity
• increasing levels of efficiency throughout the economy
'Transition' to a democratic form of government
Trends include:
• increasing transparency and accountability of government activities
• increasing levels of civic activity
Consequent environmental trends:
• dissemination of environmental information within the public domain
• independent environmental monitoring of economic activity
• government decision-making process incorporating opinions from a representative range of social constituencies
Pre-revolutionary and Socialist legacies
Examples include:
• pollution legacies (e.g. nuclear waste, toxic waste, depleted resource base)
• degraded land resources
• extensive and deteriorating technical infrastructure
• environmental management systems
• behaviours, 'ways of doing', underlying cosmologies and world views
• environmental sensibilities
• institutional inertia
BIOSPHERE
Examples of general economic trends
• marked falls in the level of industrial and agricultural production 1991-1998 (improving trends post-1998)
• service sector growth
• significant levels of unregulated economic activity
• high dependence on natural resource base and associated export markets
Associated environmental trends include:
• falling levels of gross industrial pollution output (predominantly urban based)
• limited change in the quality of air and water resources (urban)
• accentuation of certain pressures including motor vehicle pollution and domestic waste production (urban)
• unregulated appropriation of natural resources
• weak management of specially protected areas
• weakened systems of resource management
Examples of political trends
• devolution trends under Yel'tsin followed by centralising trends under Putin
• frequent change in administrative structure
• weak financing of government policies
• fractious linkages between the state and non-governmental actors
Associated environmental trends:
• weakening of environmental protection concerns within state administration
• limited success of environmental programmes
• limited productive interaction between the state and Russia's environmental movement
RUSSIAN SOCIETY
BIOSPHERE

Figure 1.1 Understanding the linkages between societal change and the wider environment

contemporary period in order to regulate the country's protected area network. Other authors have explored the way in which the establishment of a market-type economy has reconfigured the relationship between society and its natural resource base, while simultaneously indicating the need to engage fully with the specificities of change at the local level (e.g. Staddon, 2001). Herrschel and Forsyth (2001) set out to self-consciously open up debate concerning the relationship between post-socialist societal change and the environment. Operating at a general level of analysis, they touch on a number of issues which are central to the current study. In particular, they indicate the uncritical way in which the socialist environmental legacy was often conceptualised in the West during the early 1990s as well as the clear need for more local-level work. They go on to advance a critical engagement with the nature of post-socialist societal change, while at the same time remaining sensitive to sub-national variations in both the representation of, and response to, environmental issues (Herrschel and Forsyth, 2001, p. 582).

Democracy, Civil Society and the Environment

While the relationship between capitalist modes of operation and environmental processes has attracted a great deal of attention in the general literature, there has been comparatively less critical discussion concerning the links between the environment and democratic systems of governance, particularly with respect to the Russian Federation.[4] The establishment of a democratic form of government, coupled with the flowering of civil society, is often seen as a necessary prerequisite for the effective management of the surrounding environment. The increased availability of environmental information and the establishment of legal mechanisms for monitoring the activities of the state are just two of the positive aspects associated with democratic systems of governance. International organisations such as the World Bank, UNEP and the EBRD reinforce the positive environmental rhetoric associated with the democratic process (e.g. Lafferty and Meadowcroft, 1996, p. 2; see also Figure 1.1, Diagram A). Just as the failure of the command economy boosted the credibility of capitalism, similarly the failure of the Soviet authoritarian state lent considerable weight to the prevailing positive perception of the relationship between democracy and the state of the environment (Paehlke, 1996, p. 19). Nevertheless, the 'market plus civil society' development model championed in many parts of the non-Western world during the 1990s has received considerable criticism. Much of this censure has been directed towards its inability to address effectively prevailing environmental and development issues at both the local and global scale (e.g. see Fagan and Jehlička, 2003; see also McIlwaine, 1998, p. 420). More specifically, concern has been raised in relation to the vagueness of the civil society concept and the need to challenge its largely uncontested 'whiter than white' image. Superficial understandings of civil society and associated democratic structures are compounded by the reduction of such concepts to a series of easily measurable indicators (e.g. number of non-

---

[4] Pavlinek and Pickles (2000) provide some useful discussion of the tensions between democratising processes and the environment in CEE.

governmental organisations) in order to gauge the level of progress in this area (see OECD, 1999b). In addition, Western funding bodies and associated organisations can lack the flexibility or knowledge to engage with non-governmental entities deviating from prevailing Western norms of activity and organisation. For example, the professionalisation of environmental movements operating in former socialist countries, as they respond to the demands of Western funding and aid programmes, has been criticised for undermining alternative forms of indigenous public action typically more suited to the actual needs of the local community (e.g. Fagan and Jehlička, 2003; Yanitsky, 2000, pp. 63-80).

The restrictions of the nation-state paradigm, coupled with the global nature of many environmental issues, can also undermine the potential benefits associated with democratic forms of governance (see Macnaghten and Urry, 1998). For example, the complexity and uncertainty of many contemporary environmental processes militate against an effective solution via state-based democratic systems. Furthermore, the convoluted and long-term nature of environmental issues such as global warming deters the participation of the general public and opens the door to claim and counter claim, thus weakening the ability of both state and non-state actors to respond swiftly and with purpose. The evident shortcomings of state-based democracies in the face global environmental issues and transnational economic activity have led some to posit the need for establishing global democratic systems of governance and education (e.g. Holden 2002; Moiseev, 1999).

Russia, the Global Economic System, and the Environment

Conceptualising Russia's Place in the Global Economic Order

The purpose of this section is to draw attention to the wider context within which Russia's societal changes are taking place. Whereas much of the study is focussed at the national and sub-national level, processes operating at the global scale remain ever-present. In particular, the influence of global economic flows on Russia's capacity to address environmental issues is important to acknowledge. Before examining this issue in more detail, Western conceptualisations of Russia's place in the global economic order are worthy of further comment. From a general economic standpoint, the nature of the Cold War period, with its overwhelming East-West dynamic, encouraged the Soviet Union and its satellite countries to be considered under the collective label of the 'Second World'. Furthermore, there was analytical value in perceiving commonality across the region during this period with respect to the nature of the socialist system, a '"species" of social systems' according to Kornai (1992, p. 5). Since the break-up of the Soviet Union, Western conceptualisations of the region have undergone significant reworking and Russia, together with the other post-socialist states of the FSU and CEE, has been progressively incorporated into the geo-economic vision of Western regulatory organisations and their 'in-house' classification systems based on levels of income and similar macro-economic indicators (see Shaw, 1999; Smith, 2002;

2004; see also World Bank, 2003). The EBRD was established at the beginning of the 1990s in order to facilitate the movement of post-socialist countries to market-type economies. As such, it has developed an elaborate framework for evaluating economic change within CEE and FSU (e.g. see Smith, 2002, pp. 653-656). This evaluation framework centres on scoring 'progress' for each country with respect to market reforms, infrastructure capacities and the development of financial institution with success '...measured against the standards of industrialised market economies' (EBRD, 2001, p. 13). This type of approach is clearly infused with the restrictive conceptualisation of space/time outlined above whereby individual countries are assessed in relation to the development paths of 'advanced' capitalist countries; although it is conceded that the process of economic restructuring has no singular endpoint common to all. As a consequence of the application of such analytical frameworks, a marked divide has become evident in the geo-economic landscape of the former Soviet Union and its satellite countries, with CEE tending to 'outperform' those countries positioned further east. Furthermore, this process of classification along a west-east axis is grounded on the notion of a 'taken-for-granted gradation of Europeanness' the further east you travel (Kuus, 2004, p. 480). It is appropriate at this juncture to note the influence of historical factors as well as contemporary processes in helping to perpetuate such conceptualisations of space. For example, the enlargement of the EU has been effective in (re)emphasising the Europeanness of applicant countries vis-à-vis non-applicant countries such as Russia. Smith (2002) provides a critique of the unsophisticated nature of prevailing geo-economic and geo-cultural imaginations of former socialist space, and highlights the need to confront the naturalising tendencies of such categorisations in order to engage more fully with the underlying political and economic impulses behind the current reordering process. This approach has much commonality with critiques found within the general development literature concerning the way in which regions of the world are consigned to specific categories based on an array of arbitrary data. Efforts have been made to connect the two literatures in recent years (e.g. Bradshaw and Stenning, 2004; Kuus, 2004; Verdery, 2002).

Russia has limited ability to influence supranational economic processes and associated financial flows. The dramatic consequences of the 1998 global financial crash for Russia, as it rippled outwards from South East Asia, is indicative of this relative weakness. Furthermore, Russia has struggled to attract foreign direct investment (FDI) during the course of the 1990s, an ability seen by many as crucial to overall economic success (see Bradshaw and Swain (2004) for a detailed overview of FDI flows). Such a state of affairs is blamed on a host of factors including a weak financial system and high levels of domestic corruption. While it is important to dwell on Russia's vulnerability with respect to global economic processes, such an approach can also downplay other contrasting elements of Russia's global economic role. In particular, the country's immense resource base and potential for future economic growth ensure that it has substantial economic and political leverage in some quarters. For example, the enlarging EU is finding itself increasingly dependent on Russia's hydrocarbon exports and thus has a vested interest in reducing the uncertainties of Russia's economic performance over the short- to medium-term (e.g. see European Commission, 2001b).

Global Economic Processes and Environmental Concerns

An appreciation of Russia's contested position within the global economy provides a basis from which to consider associated environmental issues. The experience of other regions of the world characterised by comparatively weak national economies suggests that such frailties can give rise to specific environmental problems as core regions take advantage of weak regulation and underlying economic need in order to export pollution-intensive operations and undesirable wastes, or else undermine the integrity of local environments through the imposition of new management and technical infrastructure (e.g. see Escobar, 1995; Sachs, 1999). However, just as it is too simplistic to label Russia as an economic 'weakling', so it is insufficient to dwell only on the deleterious environmental consequences of Russia's engagement with the global capitalist system, as important as these are (see also Oldfield and Tickle, 2005). First, Russia's size and internal heterogeneity ensure that the relationship between global socio-economic processes and environmental disruption can vary considerably over the country's territory. As such, the economic and social problems of resource peripheries, such as the Russian Far East, can be contrasted with those found in the core regions of European Russia, in order to reveal markedly differing environmental concerns.[5] Second, it has been argued that countries like Russia are using Western sensitivity towards environmental issues in order to extract necessary aid and expertise to address a variety of environmental problems on their own territory (e.g. Darst, 2001). Third, and more positively, recent work concerning the ramifications of EU enlargement for the former socialist countries of CEE indicates the various ways in which environmental sensibilities can be transmitted via the functioning of both political and economic networks. In particular, Andonova (2004) draws attention to the importance of transnational industrial coalitions in facilitating the 'bottom-up' dissemination of environmental good practice and associated standards, in addition to the more obvious 'Europeanisation' of domestic environmental legislation via the demands of the *acquis communautaire*. Similarly, and momentarily laying to one side the wider conceptual concerns related to the long-term environmental viability of the capitalist system, the operations of international lending institutions and supranational organisations such as the EU, which are characterised by a rigorous process of socio-environmental assessment and evaluation, can have a positive impact on the environmental performance of small-scale indigenous economic activity by encouraging the implementation of relatively efficient production processes or more transparent operating standards. Nevertheless, the cumulative effect of lending and grant-giving activities at the regional or national scale is difficult to gauge, and can result in contradictory outcomes whereby the activities

---

[5] See Hayter et al. (2003) for a broader discussion of the particularities of resource peripheries.

of one project undermine those of another (e.g. Fagin, 2001, p. 601). This highlights the importance of scale and the fact that economic activities, which are deemed 'sustainable' at the local level may, in fact, undermine the viability of similar goals in adjacent localities. The often-ambiguous environmental consequences of economic activity, functioning at a variety of scales, underline the need for careful evaluation of the link between capitalistic activity and environmental degradation (e.g. see Forsyth, 2003, pp. 116-120). Furthermore, since revolutionary change to the functioning of the global economy is not a realistic prospect over the short- to medium-term, it would seem important to explore those elements of the prevailing system that are proving effective in generating positive environmental outcomes for Russia.

Aim of the Study

While remaining sensitive to the debates outlined above, the underlying aim of the study is to explore the way in which the recent upheaval within Russian society, precipitated by the collapse of the Soviet Union in 1991, has influenced the country's environmental situation at both the national and sub-national level. Russia's size and consequent importance for the long-term integrity of global biophysical systems provides a compelling reason for the study's engagement with the specifics of the country's contemporary environmental situation. The study provides an insight into prevailing environmental patterns and processes as they manifest themselves at different scales through a detailed exploration of environmental data and transforming administrative and legislative frameworks. At a basic level, the study consciously moves away from the often-overwhelming influence of the Soviet environmental legacy, typically couched in terms of 'disaster' and 'crisis', in order to expose the dynamism and ambiguities of the contemporary situation. This is not to sideline the Soviet experience, which remains an integral part of the overall analysis. As such, the study engages critically with the nature of the Soviet environmental legacy in order to lay the basis for a more subtle interpretation of the way in which previous institutional arrangements, systems of control and understandings of society-nature interaction are influencing, or have the potential to influence, the contemporary situation.

Structure of the Book

Chapter 2 begins by exploring the nature of the Soviet Union's environmental legacy. The first part of the chapter provides a brief overview of the negative environmental consequences of the Soviet development model and their underlying causal factors. The chapter then proceeds to examine more purposefully the extent of political and academic debate during the Soviet period prompted by the worsening environmental situation, and the corresponding responses of central state organs. The ability of Soviet society to deal with the environmental consequences of its economic excesses was clearly undermined by a host of factors. Nevertheless, a range of technical and intellectual capacities existed, which would form the basis for positive action during the 1990s.

Chapter 3 focuses on the post-1991 period and is limited to an analysis of the relationship between economic change and polluting trends within the context of Russia's macro-economic restructuring process. The extensive reorganisation of industrial, service and agricultural sectors has generated a range of contradictory environmental trends. For example, gross pollution levels have fallen substantially and further improvements are possible as new capital (both foreign and domestic) is directed towards sectors such as energy and metallurgy. At the same time, issues of waste production and motor vehicle emissions are emerging as key concerns within large urban regions. Much of the chapter explores trends at the federal level. However, the final section outlines a framework for engaging with Russia's regional environmental situation. This framework highlights the varied nature of environmental issues across the territory of the Russian Federation reflecting the country's underlying economic geography. It thus makes a strong case for the more detailed analysis of environmental trends carried out in Chapter 5.

Chapter 4 examines the development of environmental governance structures within the Russian Federation since 1991. The first part of this chapter traces the progress of legislative and policy initiatives and then proceeds to look more closely at associated restructuring within the relevant executive bodies of the Russian government. It is suggested that while significant changes have taken place in all areas, particularly during the early to mid 1990s, there remains a noticeable gap between rhetoric and practice and limited signs of any overarching policy coherence. This is the consequence of many factors and includes the strong institutional legacies of the Soviet period, weak financing, frequent organisational restructuring, the rapidity of legislative and policy development as well as deep-seated political obstacles. The discussion of state competencies is complemented by an overview of developments within the non-state sector. Russia's environmental movement, conceived very broadly, has undergone significant structural change since the late 1980s. Westernised groups tend to dominate the domestic scene but exist in tandem with a range of smaller, more parochial interest groups. In recent years, significant numbers of environmental initiatives from all over the Russian Federation have demonstrated the willingness to unite in opposition to state policies. Furthermore, the collective action of Russia's environmental organisations has also had some influence on policy development in certain areas, although this trend should not be overemphasised.

Chapter 5 provides a detailed examination of environmental trends within Russia since 1991. Importantly, this chapter does not attempt to provide an exhaustive analysis of different pollution types generated by Russia's economic activity. Instead, it focuses predominantly on key air and water pollution trends, in addition to natural resource management issues, with the purpose of establishing a general feel for the underlying state of the environment. It also builds on the regional framework provided in Chapter 3 and explores the varied nature of Russia's contemporary environmental situation. As such, the chapter is split into two distinct parts. The first part focuses on trends within urban regions and the second is concerned with environmental pressures characterising Russia's vast expanse of non-urban land (i.e. forest, agricultural regions and specially protected natural areas). The analysis reinforces the importance of resisting the certainties of orthodox (transition) conceptualisations of environmental improvement, with many urban regions remaining characterised by strained environmental situations. In addition, the restructuring of agricultural and natural resource management systems within the context of local, national and supranational processes of change, is giving rise to a range of pressures throughout significant parts of Russia's non-urban land areas, many of which demand more in-depth research. Finally, Chapter 6 re-emphasises the main findings of the study.

Methodological Approaches of the Study

Why Russia?

Before moving on to discuss the specifics of the methodology underpinning the study, it is appropriate that a few words are said concerning its explicit regional focus. Russia's noted importance for global ecological systems provides sufficient legitimisation, one might think, for the study's overall focus. However, the general environmental themes noticeable in the case of Russia have much commonality with trends evident in other countries of the FSU and CEE, and link up with more general observations concerning the 'post-socialist condition'. In this sense, there would appear much to gain from expanding the scope of the study in order to recognise and conceptualise the nature of such overlaps. At the same time, while the 'post-socialist' category continues to have analytical value at certain levels of analysis, its long-term relevance is being increasingly challenged in recognition of the complicated mixing of histories and local particularities in evidence within the countries of CEE and the FSU (e.g. Humphrey, 2002; Humphrey and Mandel, 2002). This critique implies that cross-country comparison, as valuable as it is, needs to be complemented by studies that are sensitive to the diverse range of pressures currently influencing the environmental situation, as well as environmental management and governance systems, across the post-socialist region. An obvious example here concerns the influence of the EU on the legislative base of the recent accession countries and the candidate countries of Romania and Bulgaria.

More generally, the benefits of an open and long-term commitment to a clearly defined region of study should also be acknowledged. Some of these advantages encompass practical issues. For example, language training, the development of a network of contacts in the field and the establishment of professional relationships are typically possible only within a long-term timeframe. Furthermore, a detailed focus on Russia provides the space in which to develop greater insight into the rhythms and nuances of a relatively marginalised part of the world with respect to certain sections of the Western social science literature (e.g. Willis, 2004, p. 400). Importantly, it is not being suggested that a regional focus of this nature provides superior knowledge relative to more systematic works, although it can be difficult to move away from the notions of exoticism and authority that are intimately entwined with regional specialisms (Gibson-Graham, 2004, p. 405), but rather that such knowledge is characterised by an understanding and degree of insight worthy of recognition.

Empirical Basis of the Study

In order to support the empirical focus of the study, data were collected through a combination of interviews, observational and photographic work and the analysis of secondary data sources. The bulk of the primary data collection took place between 1997 and 2001 during a number of fieldwork stays in Moscow. Further data were collected during shorter visits to Moscow over the period 2002-2004. During the course of fieldwork, interviews were conducted with a range of political actors, non-governmental representatives and academics (see Appendix 1). These interviews were particularly useful in facilitating a greater insight into the nature and scope of Russia's contemporary environmental situation as well as related administrative and policy developments. They also provided access to a variety of relevant data sources and publications. At the same time, the politicised nature of environmental issues and associated themes, combined with the request by a number of the interviewees to remain anonymous, ensures that this interview material is used largely to inform the arguments grounded in the secondary data. Secondary data analysis is based on materials published predominantly by relevant ministries and state committees. In addition, the study also incorporates a critical assessment of legislative and policy documentation, and this includes an examination of Russia's engagement with the concept of sustainable development.

The annual State of the Environment Report, compiled at present by the Ministry for Natural Resources, is a prime source of data for the study. The scope of this yearly document has broadened considerably since it first appeared at the end of the Soviet period, and the most recent edition runs into several hundred pages. Perhaps the most useful function of the report is its ability to synthesise the outputs of a range of state organisations and this includes Russia's Federal Service for Hydrometeorology and Environmental Monitoring (*Rosgidromet*), which is responsible for large-scale environmental monitoring (see Chapter 4). State reports covering health concerns, sanitary-epidemiological issues, forestry trends and natural/technical disasters are also utilised in order to obtain more specific data. The statistical reporting of the State Statistical Committee (*Goskomstat Rossii*) also forms an important element of the study. The inadequacies of these data-sets are well-rehearsed in the general literature and the figures utilised within this study concerning national and regional socio-economic and environmental trends should be interpreted with care (e.g. see Bradshaw and Vartapetov, 2003, pp. 406; Hanson and Bradshaw, 2000b; see also Chapter 3). Regional environmental data are provided in a variety of forms within the aforementioned official publications. During the 1990s, Soviet statistical conventions persisted and data were often given according to Russia's 11 so-called economic regions. These regions were utilised during the Soviet period for strategic economic planning reasons (see Pallot and Shaw, 1981, pp. 55-86). However, in 2000, President Putin, intent on reasserting centralised control over the regions, established 7 federal *okruga* to be headed by Presidential representatives, and these have now replaced the economic regions as macro-level accounting units and are referred to throughout the study (see Kistanov, 2000; Figure 1.2). The study also incorporates regional data relating to each of Russia's 89 federal units (see page *vi*). Information for the Republics of Chechnya and Ingushetia are generally missing due to the ongoing conflict and instability in that part of southern European Russia. While sub-*oblast'* (i.e. *raion*) level environmental data are available for certain phenomena, these have not been utilised within this study.

The reliance on such extensive data-sets is a consequence of the scope of the study and yet, at the same time, raises questions concerning the way in which environmental data are collected, and environmental problems conceptualised, within the Russian Federation (see also Chapter 3). At a general level, the rationale for Russia's domestic environmental monitoring network has considerable overlap with Western understandings of, and approaches to, environmental issues. In other words, Russian statistical and analytical publications divide the environment into discrete elements encompassing air, water and land resources, and anthropogenic pressures are established largely on the basis of both chemical and biological measurements incorporating a range of different compounds and known pollutants. At the same time, prevailing international conventions on environmental reporting, as evidenced by the activities of organisations such as the OECD and various bodies of the United Nations, have evidently influenced aspects of Russia's reporting procedures in order to encourage the accommodation of phenomena such as domestic waste generation as well as a more detailed analysis of the country's greenhouse gas production. The study makes use of data published by the EBRD, OECD and World Bank, although it should be recognised that in some cases these organisations obtain data directly from the relevant ministries and state committees in Moscow, and thus differ little from the data published in official reports. Nevertheless, they can provide estimates of new environmental phenomena and also reflect the findings of detailed local-level studies, which add further dimensions of understanding to the overall picture.

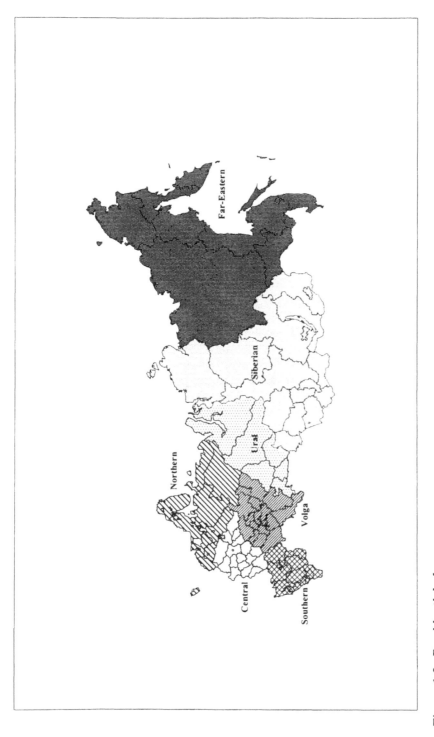

Figure 1.2 Presidential *okruga*

Chapter 2

The Soviet Environmental Legacy

Introduction

Western conceptualisations of the Soviet environment are typically coloured by the 1986 Chernobyl' accident and environmental problems of similar scale (e.g. the desiccation of the Aral Sea or the degradation of Lake Baikal). The uncertainty engendered by the Chernobyl' accident amongst the general public, both at home and abroad, together with the relative political openness of the late 1980s, was effective in creating a space for further revelations concerning the state of the Soviet environment. While the considerable extent of environmental damage and destruction is not in doubt, the confluence of so much new, and often shocking, information at this time ensured that general assessments understandably focussed on notions of disaster. This interpretation was further reinforced by the publication of 'crisis' environmental maps in official documentation and the incorporation of related concepts within Russia's legislative base during the early 1990s (e.g. Pryde, 1994). The evident willingness of the state to utilise such concepts is indicative of the acute nature of inherited environmental problems in certain regions of the Russian Federation. Western publications dealing with the environmental situation in the countries of the former Soviet Union, and made available during the early to mid 1990s, tended to focus on underlying notions of crisis with titles such as 'Troubled Lands: The legacy of Soviet environmental destruction' (Peterson, 1993), 'Ecocide in the USSR: Health and nature under siege' (Feshbach and Friendly, 1993) and 'Losing Hope: The environment and health in Russia' (Bridges and Bridges, 1996). These publications were particularly important in helping to convey to the English-speaking world both the scale and range of contemporary environmental problems inherited from the Soviet period. Furthermore, as suggested in Chapter 1, related understandings of Russia's environmental situation were internalised within debates concerning the nature of societal change as they emerged during the early years of the 1990s. Such debates tended to discredit the socialist development model in relation to the capitalist democracies of the West. It was therefore little surprise that in many instances the subsequent transformation process within Russia and other post-socialist states became associated with an improving environmental situation in addition to a host of other positive characteristics (see Chapter 3).

For some, any attempt to problematise prevailing understandings of the Soviet environmental situation as it manifested itself during the late Soviet period might appear tantamount to excusing the immense ecological damage inflicted by the socialist development model over significant proportions of the Earth's surface. However, there would seem room for a more reasoned analysis of the Soviet Union's environmental legacy in order to lay the basis for a deeper understanding of the post-socialist environmental situation within Russia. In particular, such an approach would facilitate engagement with the oft-ignored environmental sensibilities that existed behind the socialist façade and yet were evident in government and academic discussion as well as the actions of Soviet citizens during the mid-late Soviet period (e.g. DeBardeleben, 1985; Yanitsky, 2000). While it is clear that these sensibilities were able to exercise only very limited, if any, influence over the direction and effectiveness of Soviet environmental policy, they, nevertheless, provided an important pool of knowledge and 'know-how' that continued into the post-Soviet period in the form of ongoing theoretical debate, legislative activity and, in the case of certain environmental actors, political aptitude. More concrete legacies included a reasonably extensive environmental monitoring system, incorporating a range of pollution concerns, as well as a network of specially protected natural territories (see Chapter 5). An open engagement with the nuances of the Soviet period would also assist in exposing the historical richness of Russian ecological thought and encourage a more active engagement with the complexities of environmental debate within the Soviet Union as it developed during the course of the twentieth century (e.g. DeBardeleben, 1985; Oldfield and Shaw, 2002; Oldfield and Shaw, in press; Weiner, 1999). Indeed, Western conceptualisations of Russia, influenced by the powerful imagery of the Soviet period, are effective at drowning out the theoretical advances of pre-revolutionary Russian science. However, the scientific community of the nineteenth and early twentieth century was characterised by a progressive and innovative scientific understanding of the connections between society and the wider environment, as evidenced by the work of scholars such as V.V. Dokuchaev, D.N. Anuchin, P.A. Kropotkin and V.I. Vernadsky, and their approaches are worthy of engagement outside of the distorting effects of Marxist-Leninist interpretation (see also Oldfield and Shaw, forthcoming; Vucinich, 1970). This should not be seen as a purely esoteric academic exercise and the province of those interested in the history of scientific ideas alone. For example, the relevance of Vernadsky's work on the biosphere, which was published in the 1920s, for contemporary Western understandings of global ecological systems has been noted by a number of authors (e.g. Grinevald, 1998). Furthermore, the conceptual framework established by these scientists remains influential in contemporary Russia, most notably in relation to understandings of sustainable development (see Chapter 4; Oldfield and Shaw, in press).

In the light of the previous discussion, this chapter is divided into four main parts. First, the environmental situation at the end of the Soviet period is highlighted based on an analysis of both official documentation and relevant Western publications. This overview retains the thematic division between air, water and soil resources utilised in the relevant Soviet publications. It also draws attention to the acute nature of the environmental situation that prevailed in a number of regions as well as more generalised trends related to urban air quality and the state of water resources. Second, the underlying factors behind the emergence of Russia's strained environmental situation are explored in more detail with the environmental shortcomings of the Soviet development model forming a main focal point. Third, the responsiveness of the Soviet political bureaucracy to emerging environmental problems is examined via an assessment of policy and administrative initiatives. Fourth, civic expressions of environmental concern during the mid-late Soviet period are also explored.

Overview of Environmental Trends at the End of the Soviet Period

Introduction

The first State of the Environment Report for the USSR was published in 1988 and represented a visible sign that the administrative reorganisation of the late Soviet period was having at least some positive impact addressing environmental concerns. An abridged English version of this report was published later in 1990 and this drew attention to the Soviet Union's 'tense' ecological situation, although at the same time noting the regionally varied character of the underlying trends (see Goskompriroda, 1990, p. 4). Zeev Wolfson, writing in the foreword to Pryde's *'Environmental management in the Soviet Union'* at the beginning of the 1990s, noted that:

> ...a book about ecological problems in the Soviet Union would be timely at any point in the last twenty to thirty years for the simple reason that environmental disasters developing on a territory equal to one-sixth of the entire world land mass cannot help but influence the ecological balance of the entire planet (Wolfson, 1991, p. xv).

Similarly, a joint report concerned with the Soviet economy compiled by a number of international economic and financial organisations posited that:

> The magnitude of environmental problems in the USSR cannot be estimated with precision: the information is partial and in some cases it has been distorted because of bureaucratic interests. Nevertheless, sufficient information exists to conclude that many of the industrial and agricultural regions are on the verge of ecological breakdown, posing an imminent threat to the health of present and future generations (IMF et al., 1991, p. 1).

The three publications differ in their allusion to the scale of the Soviet Union's prevailing environmental problems, and yet all are indicative of an underlying sense of urgency to address longstanding problems.

(Photograph: courtesy of John Cole)

Plate 2.1 Scenes of industrial activity in the urban region of Zheleznogorsk (southern European Russia), early 1980s

General Environmental Trends

By the end of the 1980s, the Soviet Union's total volume of annual air pollution emissions was approximately 100 million tonnes, which according to Bond et al. (1990, p. 404) represented roughly 75 percent of the USA total. At the same time, a comparison based solely on gross output figures hides the underlying production inefficiencies of the Soviet economy relative to Western economies during this period. Similar disparities were noted with regard to the Soviet Union's motor vehicle emissions. Thus, while numbers of vehicles were minimal in comparison with the USA, the recorded level of air pollution from this source was more than 60 percent of the US total (Bond et al., 1990, pp. 404, 406). The Soviet economy's high level of pollution intensity (i.e. the volume of industrial production output correlated with the volume emissions of main pollutants) relative to the West remains a defining feature of Russia's environmental situation in the contemporary period (see Chapters 3 for a more detailed analysis). While such comparisons are useful in highlighting the relative inefficiencies of the Soviet production system, it should also be noted that the volume of air pollution emissions (this refers to both stationary source and mobile source air pollution) generated by the Soviet Union fell steadily throughout the 1980s (Kostenchuk et al., 1993, section 2.1.1) and this was a trend repeated throughout many of the former socialist countries of CEE (e.g. see Pavlinek and Pickles, 2000). The decline during this period was related strongly to a combination of fuel switching policies, with gas becoming more dominant within the domestic economy, as well as associated technological improvements in the production process (see Peterson, 1993, pp. 45-49). While it is important not to overplay such trends, they are nevertheless suggestive of the Soviet Union's concern to address environmental problems in a purposeful manner. Further evidence of this is found in the growth of the Soviet Union's environmental monitoring system (see Chapter 5). By the end of the 1980s, the network of air quality monitoring stations incorporated the majority of urban areas with populations of more than 100,000 in addition to those smaller settlements characterised by significant industrial activities. The main pollutants forming the heart of the monitoring system included suspended substances, nitrogen oxides (NOx), sulphur dioxide (SO_2), and carbon monoxide (CO) with additional pollutants included on a more selective basis (Kostenchuk et al., 1993, section 2.1.1). It is clear that a significant number of urban regions within the former Soviet Union suffered from marked levels of air pollution at the end of the 1980s (Goskomstat SSSR, 1989, pp. 14-31; Peterson, 1993; Pryde, 1991). More specifically, the 1991 State of the Environment Report for the Russian Federation described levels of urban air pollution in Russia as 'high' and linked the most significant problems to main sectors of the economy such as metallurgy, fertilizer production and the chemical/petro-chemical branches (MinEkologiya, 1992, p. 9). The report also indicated that some urban regions had registered stabilising or improving levels of air quality during the period 1986 to 1991, although the situation remained strained in other cases (MinEkologiya, 1992, p. 10). Furthermore, the report noted that 84 urban localities (approximately 25 percent of urban regions incorporated within the monitoring network) had recorded levels of air pollution at least 10 times the permissible limit for one of more pollutant during the course of 1991. This pessimistic assessment was supported by sanitary-epidemiological measurements of air quality published in the 1992 report concerning the State of the Population's Health (MinZdrav, 1994, p. 3).

In contrast to air pollution, there was no marked decline in water pollution discharges during the late Soviet period. Furthermore, water quality was a significant issue for a large proportion of the USSR's river and reservoir network (e.g. Kostenchuk, 1993, section 2.3; MinZdrav, 1994, pp. 4-5). The 1988 State of the Environment Report for the Soviet Union indicated that the Dniestr (south-west Ukraine and Moldova) and Don (southern European Russia) together with rivers on Sakhalin Island and the Kola Peninsula were amongst the most polluted in the Soviet Union (Goskompriroda, 1990, p. 15; see Figure 2.1). More generally, a substantial number of water sources registered levels of certain pollutants in excess of permissible levels. Importantly, water pollution problems were not related exclusively to industrial sources but included substantial contributions from the municipal and, to a lesser extent, agricultural sectors of the economy.

By the late 1980s significant areas of the former Soviet Union were suffering from land degradation in its various forms. While ameliorative efforts were evident, levels of soil erosion, salinisation and loss of soil fertility were substantial (Pryde, 1991, pp. 198-202). Land resources were also subjected to disruption from the extensive development of mineral resources, although, as Bond et al. (1990, p. 426) indicated, there had been a move towards more intensive production practices during the 1980s. Waste issues were also recognised as being problematic at this time. Industrial enterprises were responsible for their own waste and this had resulted in the accumulation of substantial volumes of toxic waste on industrial sites. The Soviet Union had a poor record with regard to the safe storage and treatment of nuclear waste, with weak regulation traceable to the early years of the Soviet nuclear programme (e.g. see Peterson, 1993, pp. 140-153). For the most part, municipal waste was disposed of in landfill sites, although incineration was the preferred method in some urban regions of the former Soviet Union (Bond et al., 1990, p. 434; Peterson, 1993, 125-136). The volume of municipal waste produced by Russia at the end of the Soviet period was relatively low in comparison with Western economies. For example, OECD figures suggest that while the average Russian citizen produced 190kg of municipal waste in 1990, citizens in the USA were producing over 600kg of waste (OECD, 2002b, p. 11).

Environmental Mapping

The introduction to this chapter made reference to the development of environmental maps during the late 1980s and early 1990s, which were then used to facilitate a greater understanding of the scale and scope of environmental problems throughout the region. The Institute of Geography, Academy of Sciences USSR, was instrumental in advancing work in this area (see Kochurov, 1989; Peterson, 1993, pp. 7-11; Pryde, 1994). According to maps published at the end of the Soviet period, approximately 16 per cent (3.7 million km^2) of the Soviet Union's territory was classified as having a 'critical' (*ostryi*) environmental situation, and this included regions characterised by 'crisis' (*krizisnyi*) and 'catastrophic' (*katastroficheskii*) conditions (Kochurov, 1989, pp. 13-14, 16). A 'crisis' designation indicated significant but reversible damage inflicted on the highlighted area whereas a 'catastrophic' designation was indicative of irreversible damage. Kochurov (1989, p. 16) noted that by the end of the 1980s, over 25 percent of the Soviet Union's population lived in urban regions situated within areas characterised by an 'acute' environmental situation. More generally, regions exhibiting a strained environmental situation were located throughout the Soviet Union in Central Asia, the Caucasus, the Slavic Republics of Russia, Belarus, and Ukraine as well as the Republic of Moldova. Reasons for the 'crisis' and 'catastrophic' designations ranged from high levels of air and water pollution (largely urban based) to nuclear contamination and soil damage.

Similar mapping work carried out by the Institute of Geography, but this time specific to the Russian Federation, was incorporated within the country's 1992 State of the Environment Report. This suggested that approximately 15 percent (2.5 million km^2) of Russia's territory was characterised by a 'critical' (*ostryi*) or 'severely critical' (*ochen' ostryi*) environmental situation. In this case, the latter designation implied that human health was at serious risk from the prevailing environmental situation. Thirteen more or less discrete regions within the Russian Federation were considered representative of a 'severely critical'

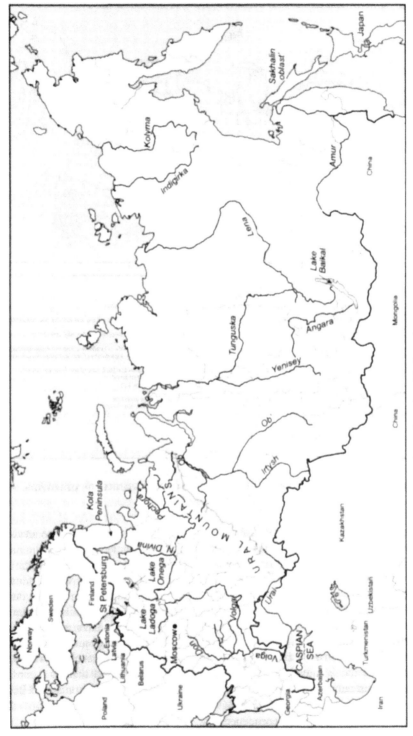

Figure 2.1 Selected physical features of the Russian Federation

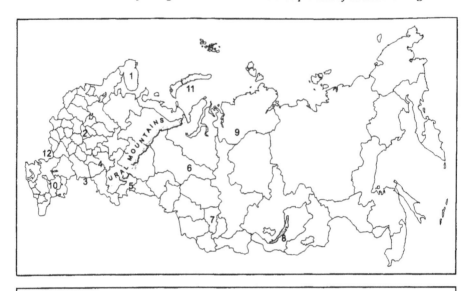

Figure 2.2   Regions characterised by severe environmental problems, early 1990s

ecological situation (see Figures 2.2 and 2.3). Most of these regions were located in the more densely populated European part of Russia and included parts of the middle Volga and southern Urals industrial zones. The four regions located east of the Urals were restricted to parts of east and west Siberia. The island of Novaya Zemlya, located off the north coast of Russia in the Arctic ocean, was a key site for nuclear tests during the Soviet period and remains characterised by significant levels of radioactive pollution. It is difficult to determine the precise area affected by the Chernobyl' explosion and Figure 2.3 is based on the percentage area of agricultural land, within a given federal unit, polluted by Caesium-137. As such, it provides an indication of the fallout path from the reactor with the expected highest recorded levels of contamination located on Russia's western border with Ukraine and Belarus (see the work of Marples (e.g. 2004) for a more detailed examination of the Chernobyl' explosion and its consequences).

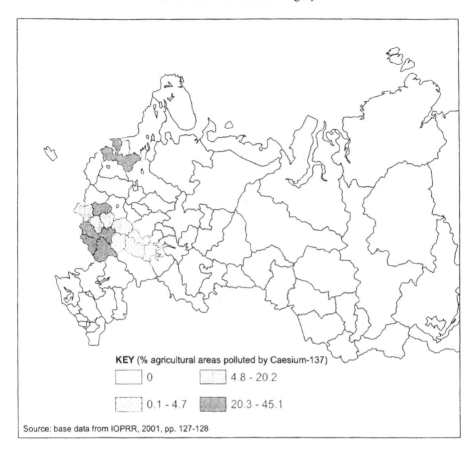

KEY (% agricultural areas polluted by Caesium-137)

☐ 0		▨ 4.8 - 20.2	
▨ 0.1 - 4.7		▓ 20.3 - 45.1	

Source: base data from IOPRR, 2001, pp. 127-128

Figure 2.3   Agricultural land affected by Chernobyl' fallout, late 1990s

The centres of Soviet industrial activity were obviously instrumental in determining the geography of regions characterised by serious environmental degradation. In addition, agricultural activity was responsible for encouraging high levels of soil erosion, particularly in the Republic of Kalmykia (North Caucasus region). Pryde (1994, p. 46) highlights the linkages between this mapping work and articles 58 and 59 of the law 'Concerning the Protection of the Natural Environment' (passed in 1991; see Chapter 4 for a more general discussion of Russian environmental law). In particular, these two articles introduced the concepts of 'ecological emergency zones' (*zony chrezvychainoi ekologicheskoi situatsii*) and 'ecological disaster zones' (*zony ekologicheskogo bedstviya*). These were official designations and once applied to a region allowed for the channelling of necessary funds to facilitate rehabilitation work (Pryde, 1994, p. 47). According to the aforementioned 1991 environmental law, ecological emergency zones denoted regions where the environmental quality was such that it threatened the state of the natural ecological system and thus posed a substantial danger to public health. In contrast, ecological disaster zones referred to areas where economic or other activities had resulted in a deep and irreversible change to the natural environment (see also Bond and Sagers, 1992, p. 470). The concepts of ecological emergency and ecological disaster zones have been retained in the revised environmental protection law issued in 2002 (article 57).

It is clear from this brief overview that the Soviet Union was characterised by a range of marked environmental problems during its final years. Unsurprisingly, these problems were correlated strongly with areas of industrial activity and urban regions. The following section explores in more detail the main reasons behind the emergence of the Soviet Union's environmental situation.

The Soviet Production of Nature

Soviet Industrialisation and the Transformation of Nature

The environmental problems of the Soviet period are routinely attributed to a number of prevailing characteristics of the Soviet development model together with its underlying ideological basis. Foremost amongst these is the Soviet Union's founding on modernist ideals, which afforded considerable emphasis to material development based on extensive production techniques and the application of advanced technology in areas such as nuclear power, space travel, and military weapons. This basic belief was reinforced by an ideologically-driven need to prove the superiority of the socialist system vis-à-vis the capitalist West together with a deep mistrust of the non-socialist world. As such, the relative economic backwardness of the Soviet Union in comparison to countries of the capitalist West spurred a truly enormous industrialisation drive from the 1930s onwards creating the basis for many of the environmental problems that would follow.

The 1920s in the Soviet Union were years of marked contrast. The internal strife following the 1917 revolution undermined the development of the Soviet Union's economic base and both created and aggravated a host of social problems. At the same time, it was a period characterised by intense intellectual and creative activity as some of society's brightest individuals responded energetically to the challenge of creating a better future for the Soviet people. This was perhaps most obvious in artistic circles spearheaded by the key figures of the *avant-garde* artistic movement. Arran Gare (1996) suggests that similar inventiveness was also a feature of ecological thought during the 1920s citing as evidence the work of individuals such as A.A. Bogdanov and the ecologist V.V. Stanchinskii. Nevertheless, the rise of Stalin during the late 1920s curtailed these intellectual developments in much the same way that it blunted a plethora of other nascent social initiatives. It was at this time that Soviet policies towards the wider environment began to resonate most clearly with the now familiar tenets of human domination over nature and when the notion of 'building socialism' was to inspire a generation (Fitzpatrick, 1999, pp. 67-71; Weiner, 1988, pp. 168-171). The confrontation between society and nature was effectively individualised by the 'heroic' spirit of the 1930s, which alerted the public to the exploits of Soviet men and women (Fitzpatrick, 1999, pp. 71-75). Such heroism comprised personal courage in tandem with technological advancement, a combination that was very evident in the case of Soviet polar exploration (Fitzpatrick, 1999; McCannon, 1995). As McCannon (1995, pp. 16, 21) suggests, the Arctic was easily portrayed as a suitable natural element against which to test the capacities and strength of Soviet society. The enforcement of the First Five-Year Plan[1] at the end of the 1920s ensured that industrial expansion became an integral part of the Soviet development model and, whilst there were many problems and inefficiencies, the Soviet Union recorded some remarkable increases in gross output levels during the course of the following decade (Nove, 1992, pp. 231-233). These increases in production resulted in significant areas of land being developed for industrial purposes and created an industrial geography that would play a defining role in the Soviet Union's later environmental problems.

Soviet industrial development, however, did not take place in a vacuum but built upon the pre-revolutionary patterns of industrial production and it is worth dwelling momentarily on these. By 1914, the Russian empire covered a vast area of land stretching all the way from eastern Europe to the Pacific coast (e.g. see Lieven, 2000). While production levels of key industrial goods such as iron and steel lagged behind those of the USA, UK and Germany, they nevertheless were significant at a European scale (Nove, 1992, pp. 1-8). At this time, much of Russia's industrial activity was located west of the Urals with significant hearths of production situated in the Donbass region (predominantly Ukraine), the urban agglomeration centred on Moscow and the southern Urals (e.g. see Shaw, 1999, p. 29). In addition, the hydrocarbon resources of the Caspian Sea basin provided a southern outpost of industrial activity centred on the Azerbaijani administrative center of Baku (see O'Hara, 2004). Stalin's industrial drive reinforced these pre-revolutionary industrial strongholds whilst developing new industrial centres further east for both strategic and developmental reasons. The main locations of this eastward expansion included the Volga basin as well as southern parts of the Urals and Siberia (see North & Shaw, 1995; Pallot & Shaw, 1981; Shaw, 1985; 1999). The evolution of the Soviet Union's industrial base continued in the post-war period with the notable development of hydrocarbon resources in the north-west regions of Siberia. It is difficult to convey the enormity of the industrial drive during the Stalin years. Kotkin's study of the construction of an integrated iron and steel complex at Magnitogorsk (southern Urals) during the 1930s is effective in providing an insight into the breadth of Soviet vision at this time as well as the ugly realities associated with the construction of Soviet society. Despite high levels of disorganisation and wastefulness, by the end of the decade Magnitogorsk was producing over 5 million tons of iron ore, nearly 2 million tons of coke and approximately 1.5 millions tons of steel (Kotkin, 1995, p. 62). In order to support such massive production levels, a similarly large settlement was established during the 1930s with an official population of almost 150,000 by 1939 (Kotkin, 1995, p. 130). The physical achievements at Magnitogorsk were repeated to a greater or

---

[1] The 'optimal variant' of the plan was introduced in 1929 and reportedly completed ahead of schedule in 1932 (e.g. see Nove, 1992, pp. 143-146).

lesser extent in various parts of the Soviet Union as new centres of industrial production were opened up.

The industrial drive under Stalin was accompanied by correspondingly significant environmental projects aimed at facilitating the economic transformation of the country. Comparison can be drawn between the scope of these environmental projects and similar large-scale activity in the USA during the depression years, epitomised by the construction of the Hoover dam on the Colorado river (e.g. McNeill, 2000, pp. 157-159). In the Soviet Union, water resources were managed for hydro-electric energy production, navigation, and irrigation purposes from the 1930s onwards. This resulted in the establishment of an extensive reservoir network and the subsequent flooding of large areas of land as well as the disruption of river regimes in European Russia and beyond. According to Pryde, the creation of the Kuybyshev reservoir during the 1950s, which is located in the middle Volga region, resulted in the greatest loss of land with approximately 280,000 hectares of agricultural land disappearing under water (Pryde, 1972, p. 115). Flooding problems associated with the reservoir construction of this period were noted during the late Soviet period (Kostenchuk et al., 1993, section 2.1). While such activities ensured the loss of significant chunks of agricultural land, entirely separate schemes were outlined in order to put hitherto marginal land under the plough. The Stalin Plan for the Transformation of Nature (initially outlined in 1948) was developed with the intention of altering the local climate of the southern steppe region through the planting of extensive shelterbelts and the creation of irrigation networks in order to facilitate increases in the level of agricultural production (Bater, 1989, pp. 134-136; Matley, 1966, p. 102; Pryde, 1972, pp. 16-17; Weiner, 1999, pp. 88-93). Khrushchev's Virgin Lands Scheme has also received considerable attention in the Western literature with its attempt to utilise the agricultural potential of northern Kazakhstan and the southern regions of West Siberia (e.g. Bater, 1989, p. 136). River diversion projects were touted in order to address, amongst other things, the irrigation needs of both southern European Russia and the Central Asian Republics (e.g. see Weiner, 1999, pp. 414-415). The proposed diversion of the river Ob' and the river Irtysh (see Figure 2.1), to enable approximately 30 cubic kilometres of water to be transported more than 2 thousand kilometres southwards to the Aral Sea region, would have resulted in a marked change in the hydrological regime of western Siberia if it had been implemented (Micklin, 1991, pp. 227-230; Weiner, 1999, pp. 415). While this particular initiative was not realised, the scope of such environmental projects provides an insight into the potential capacity of Soviet society to transform nature and the corresponding faith placed in the technical capacities of the Soviet system and its citizens by the authorities.

While the theme of transforming nature was tempered to some extent by later concessions to the possibility of nature having some influence on societal development (e.g. Matley, 1966, p. 107; see also McCannon, 1995), it nevertheless remained a defining characteristic of the Soviet period. Furthermore, the nature-society confrontation ensured that new technologies, and especially large-scale industrial technologies, were enthusiastically embraced by the Soviet regime, particularly where they promised superiority in relation to the non-socialist world.

The case of nuclear power provides a useful illustration of this. The Soviet nuclear program was driven primarily by military concerns (see Holloway, 1994), and yet peaceful applications of nuclear energy were also sought and deployed. Domestic energy demands had been a cause of concern during the infancy of the Soviet Union and Lenin's vision of socialist development had been associated closely with the expansion of the country's electricity generating potential. Nevertheless, it is important to recognise that whilst the potential of nuclear power certainly appealed to the mindset of Soviet planners, there was nothing inevitable about its use in the Soviet domestic context. Rather, the emergence of the Soviet Union's domestic nuclear potential required concerted promotion by its main protagonists, men such as Igor' Vasil'evich Kurchatov (Josephson, 1999, pp. 6-9). In a similar vein, Holloway (1994, pp. 364-365) cautions against the acceptance of an uncritical and deterministic association between the Soviet system and its nuclear program and instead draws attention to the international context within which the development of the first nuclear bomb took place. Nevertheless, he goes on to suggest that the development of nuclear technology also appealed to a regime intent on challenging and eventually superceding the West. The military and peaceful outcomes of the Soviet Union's nuclear programme were extensive and included the establishment of closed nuclear cities, a fleet of nuclear-powered submarines and icebreakers in addition to the detonation of a significant number of 'peaceful' nuclear explosions for industrial and related purposes (Josephson, 1999, pp. 297-308). Furthermore, by 1990, Russian atomic power stations had an electricity generating capacity of 20.2 million kilowatts, which represented approximately 9.5 percent of total generating capacity (Goskomstat, 2000, p. 147; see also Chapter 3). These achievements must be balanced against the frequent instances of radioactive leakage, wholly inadequate waste handling facilities (which included the dumping of radioactive waste at sea), in addition to major accidents, most notably the 1957 Kyshtym explosion in the Urals region (see Pryde, 1991, pp. 99-100) and the Chernobyl' explosion in Ukraine. Josephson (1999) provides a damning evaluation of the achievements associated with the Soviet nuclear power program, highlighting the industry's blind faith in technological advancement coupled with the downplaying of public health issues. At the same time, the experiences of the Soviet Union should not be considered in isolation. For example, it is instructive to contrast the Soviet situation with the legacy of public health and environmental problems associated with the US nuclear program (both military and domestic) with instances of near catastrophe (Three Mile Island) and contamination evident during the course of last fifty years (e.g. see Dalton et al., 1999).

Environmental Weaknesses of the Planned Economy

As indicated above, various aspects of the Soviet system and its underlying ideology of Marxism-Leninism have attracted environmental critique within the Western literature. These include the incentive structure of the production system itself in addition to the innate assumption that a centrally planned economy would be more environmentally sound that its capitalist counterpart thus leading to limited self-critique (e.g. DeBardeleben, 1985, p. 151). In relation to the first point, the Soviet system was geared towards rewarding the fulfilment of planned targets situated within a general framework of extensive development and increasing physical output. The centralised administration operated a 'taut' planning system whereby planned production targets deliberately overestimated available resource inputs and this engendered a production system obsessed with the achievement of multiple production targets under conditions of supply shortage (Åhlander, 1994, p. 38; Kornai, 1992, p. 126). Furthermore, the hierarchical production structure placed considerable pressure on those individual enterprises at the bottom of the chain of command in order to meet the set targets of the various ministries and middle-agents (e.g. Kornai, 1992, pp. 97-100, 110-130). Under such conditions, environmental concerns were understandably secondary to levels of gross production and resources for reducing pollution output were difficult to capture (see Åhlander, 1994). Indeed, there is considerable anecdotal evidence to suggest that under conditions of production maximisation and supply shortage enterprise managers would disable environmental cleaning equipment and generally cut corners in order to ensure plan targets were met (e.g. Goldman, 1972, pp. 63-70; Peterson, 1993). In addition, the prevailing accounting system was based largely on material indicators and thus there was no reason for the enterprise manager to adopt a more environmentally informed stance. The 'top-down' economic planning structure dominated by the narrow interests of industrial production ministries has received much critical attention over the years from Western observers. For example, DeBardeleben (1985, pp. 218-219) highlights the inherent conflict internalised within such a system between the overwhelming incentive to focus on narrow production concerns and targets and the competing need for regional planning in order to address operational inefficiencies and emerging environmental problems at the local level. While regional administrative bodies existed during the Soviet period, these had limited power to challenge the planning directives of industrial ministries. There were periods of experimentation, most notably under Khrushchev, and later with the development of the territorial production complexes, which attempted to facilitate the complex interaction of various ministries in a specific region (see DeBardeleben, 1985, pp. 221, 225-226; Shaw, 1999, p. 77).

The Soviet system thus tended to encourage an 'output at all costs' mentality where under-production was considered a major failing and over-production positively rewarded. The increasing complexity of the Soviet economic system as it developed placed considerable strain on central planning structures and resulted in oversights and further shortages within the production system. These inefficiencies were compensated for partially by the emergence and consolidation of unofficial linkages between production units. Such linkages were typically coordinated at the local level in order to transgress the rigid vertical production lines imposed from above. The emergence of informal connections in order to negotiate perceived and actual inefficiencies extant within the system was not restricted to production units but was also present more widely in Soviet society (e.g. Fitzpatrick, 1999, pp. 62-65; Ledeneva, 1998; Nove, 1992, pp. 202-

203). It is important to recognise that the environmental weaknesses of the Soviet system were not simply discussed in the West, but also formed the basis of intensive debate within Soviet political and academic circles from the late 1960s and early 1970s. The work of DeBardeleben (1985) is particularly valuable here in drawing attention to the nature of such debates and the growing interest in issues such as improving the efficiency of natural resource use as well as more general concerns related to the ecologisation of production processes. The following section provides further insight into the attempts by the Soviet state to address the country's growing environmental problems.

The Responsiveness of the Soviet State to Environmental Issues

Environmental regulation, broadly defined, existed in various forms during the Soviet period and is also traceable to the pre-revolutionary era in Russia. For example, the regulatory activities of Peter the Great concerning natural resources have been highlighted by authors such as Pryde (1972, pp. 9-10). The dynamics of state-led environmental protection activity during the early Soviet period were both fluid and complex and are beyond the scope of this chapter. For an excellent discussion see the work of Douglas Weiner (1988; 1999). At the same time, it is important to acknowledge the attempts made by the nascent Party administration to address conservation issues at this time. Writing in 1987, Lisitzin (1987, pp. 314-316) identified three distinct periods of legislative development during the Soviet period geared towards addressing conservation and related environmental concerns. The first encompassed efforts to enforce nature protection laws during the regime's first three decades. The second period spanned the 1950s-1970s and was characterised by the introduction of more sophisticated nature protection legislation at republic level and the 'greening' of the legal system via the internalisation of environmental norms and standards. The third period was underscored by the environmental concessions evident in the revised 1977 Soviet Constitution (see also Pryde, 1991, p. 6; and Chapter 4). This categorisation is necessarily arbitrary and more recent attempts to assess the evolution of environmental legislation and policy during the Soviet period understandably differ in their interpretation of underlying themes and key events. For example, an official publication of the Russian Federation's Ministry for Environment Protection and Natural Resources identified 1972 as a key date heralding a new stage in the development of the USSR's environmental legislation (Kostenchuk et al., 1993). This date coincides with the UN Conference on the Human Environment held in Stockholm and, while the Soviet Union was not an active participant, the event emerged as a marked watershed in the development of a global environmental consciousness. As such, the outcome of the conference had an impact on Soviet environmental policy and associated thinking. In addition, 1972 also witnessed the publication of a domestic decree 'Concerning Measures for the Further Improvement of Nature Protection and the Rational Utilisation of Natural Resources' (Kostenchuk, 1993, section 4.2.1). More generally, branches of law dealing with water, mineral and forest resources were augmented during the course of the decade. There was a further flurry of relevant legislative activity towards the end of the Soviet period and a key outcome of this was the aforementioned 1991 RSFSR law 'Concerning the Protection of the Natural Environment', which was enforced shortly before the demise of the Soviet Union. A report by the OECD referred to it as an 'umbrella' law acknowledging its potential to provide the basis for the development of a range of more specific environmental laws (OECD, 1996, p. 18). Importantly, this law provided a main steer for the development of Russia's environmental legislation and associated policy during the 1990s (see Chapter 4). Mazurov's (2004) recent interpretation of Russian environmental policy development provides further insight and identifies five distinct periods. The first extends from the late nineteenth century to the 1960s and is characterised by the implementation of various conservation and environmental protection measures. The second period (1960s–1972) is defined by the emergence of a more rigorous conceptual foundation underpinning the formation of environmental policy and coincides to a large extent with the observations of Lisitzin outlined above. The intellectual debate sparked by the 1972 Stockholm Conference forms the basis of the third distinct period (1972-1988). More specifically, this phase is associated with the development of environmental regulation infrastructure within the context of a planned economic system. This is contrasted with the following phase (1988-2000), which is characterised by a concerted effort to establish market-based mechanisms in order to regulate pollution emissions and resource-use in the aftermath of the Soviet system's failure. A final phase (post-2000) highlights the evident shortcomings of Russian environmental policy and the need for a more scientifically rigorous and focused management infrastructure (see Chapter 4).

A consistent criticism of the Soviet environmental management system concerned the lack of a centralised, independent state body responsible for environmental issues, a situation that was to remain unchanged until the final years of the regime (Pryde, 1991, p. 10). At the same time, it should be noted that a number of Union Republics established their own republic-level bodies during the Soviet period (Lisitzin, 1987, p. 318). Nevertheless, at All-Union level, environmental competencies and responsibilities were dispersed throughout the administrative system, which undermined the establishment of a coherent and effective policy base. The situation was made worse by the noted tendency to plan the economy along vertical ('top-down') and departmental lines rather than on a regional basis. Leadership in the area of nature protection and resource-use resided with the USSR Supreme Soviet and the Council of Ministers. These bodies were supported by the State Planning Committee (*Gosplan*) which was responsible for planning issues related to environmental management as well as the rational use of natural resources. Beyond these administrative bodies, the responsibility for implementing environmental policy was devolved to the relevant ministry (see Lisitzin, 1987, pp. 316-318). Obviously, the delegation of both resource-use/processing and environmental protection functions to individual ministries compromised the effectiveness of the latter, with economic concerns and plan fulfilment taking precedence as already noted above. Furthermore, the wide dispersal of environmental functions within the administrative system ensured that the coordination of associated policy initiatives was difficult to ensure. These weaknesses were acknowledged by the Soviet administration in the late 1980s and resulted in the 1988 decree 'Concerning the Radical Restructuring of the Country's Nature Protection Affairs' which established the State Committee for the Protection of Nature (*Goskompriroda SSSR*). *Goskompriroda SSSR* was given responsibility for issues of environmental protection and natural resource use in tandem with republican level Councils of Ministers (Goskompriroda, 1990, p. 38; see also DeBardeleben, 1990, p. 251). In spite of this positive development, DeBardeleben (1990, p. 252) noted a range of weaknesses characterising the newly formed body and these included inappropriately trained staff as well as unclear lines of responsibility, criticisms which were to remain relevant for the post-1991 Russian environmental management infrastructure (see also Peterson, 1995b; Chapter 4).

Soviet Society and the Environment: Science, Civic Action and Protest

The nature of the Soviet totalitarian system ensured that critical scholarly debate was an activity fraught with difficulty and while the environmental inadequacies of the Soviet planning system stimulated academic discussion in addition to some concrete improvements, the role of the Party itself was rarely questioned. Nevertheless, there were individuals prepared to test the boundaries of accepted practice and prevailing ideology. For example, with respect to the debate concerning nature-society interaction, the case of the geographer V.A. Anuchin is particularly interesting. In his 1960 book 'Theoretical problems of Geography', Anuchin advanced a more developed understanding of the relationship between nature and human society than the one decreed by Stalin (Matley, 1966, p. 105). In particular, he suggested that human society should be considered an integral element of the geographical environment thus transcending the rigid separation between the two spheres prevalent at the time and simultaneously advancing an acknowledgement of nature's potential influence on the development of human society. Anuchin struggled against the inherent conservatism of Soviet academia (see Hooson, 1962) in order to advance his ideas and, indeed, modified his views over time in apparent response to some of this criticism (Matley, 1966, p. 108). Nevertheless, it should be noted that the official Soviet stance on human-nature interaction was widened after Stalin's death in order to acknowledge the likelihood of nature exerting some influence over the development of Soviet society.

More generally, academic debate, reflecting on a growing awareness of emerging environmental problems, concerned itself with a number of key issues during the mid to late Soviet period such as the incentive structure of the production system, underlying bureaucratic inefficiencies and a critique of the Marxist labour theory of value (e.g. DeBardeleben, 1985, pp. 241-265). Debates related to the compatibility of economic growth and environmental protection have some overlap with the more recent debates concerning sustainable development in the West. The depth and complexity of such debates as they emerged during the Soviet period, in addition to their similarity in many instances to analogous

Western discussions, is often absent from environmental critiques of the period. Indeed, the theoretical arguments of Anuchin concerning nature-society interaction coupled with the aforementioned innovative work of Soviet scientists in areas of ecology and related disciplines are indicative of the intellectual discussion that persisted during the Soviet period in spite of the restrictive nature of the Soviet regime. Furthermore, it is important to recognise that while Stalinist purges undermined severely intellectual linkages with the pre-revolutionary period, these connections were by no means destroyed. As such, the innovative work of scholars such as V.V. Dokuchaev (1846-1903), A.I. Voeikov (1843-1916) and V.I. Vernadsky (1863-1945), in relation to nature-society interaction, remained influential during the Soviet period (e.g. see Bailes, 1990). As mentioned in the introduction, the ideas of Vernadsky related to the biosphere continue to form the basis for considerable discussion in the contemporary period (See Chapter 4; Oldfield and Shaw, 2002; in press).

The value of open discussion and public protest in addressing environmental problems also received limited acknowledgement from central state organs of power during the Soviet period. While relatively significant volumes of environmental data were generated during the mid-late Soviet period, as testimony to the regime's growing concern over the state of the environment, these were not released to the public at large. Nevertheless, environmental concern did move on occasion beyond the confines of academic and political debate in order to actively oppose the Soviet state (e.g. Pryde, 1991, pp. 246-250). Weiner (1988; 1999) outlines a detailed and convincing argument for the existence and resilience of a core of environmentally-minded scientists during the Soviet period. Importantly, the sensibilities of this group were strongly defined by conservation concerns with their intellectual roots traceable to the pre-revolutionary period (see also Oldfield and Shaw, forthcoming). Furthermore, key strands of this conservation ethic were imbued with practical concerns to understand more fully the functioning of natural systems far removed from the influence of human society in order to facilitate the development of efficient agricultural and related systems (see also Gare, 1996). This was underpinned by a belief in the existence of definable, self-regulating natural units (biocenoses) within the biosphere and provided the conceptual basis for the network of *zapovedniki* (nature reserves), which placed a premium on preserving distinctive natural units for scientific study (Weiner, 1988; 1999). It would seem that the Soviet authorities were sufficiently unperturbed by the dissent of this conservation movement to tolerate its existence even while ruthlessly undermining the agency of other civic initiatives. The environmental debate within Soviet society widened during the 1960s with respect to both its intellectual sweep and social constituency. While preservation and conservation concerns grounded on an increasingly questionable model of nature remained active, they were subsequently joined by other interest groups concerned with environmental degradation and the loss of cultural heritage (e.g. Weiner, 1999, p. 356). The latter concern was prompted, at least in part, by the marked loss of land resulting from hydroelectric developments and similar large environmental schemes mentioned above. A number of scholars have attempted to trace the development of environmental sensibilities within Soviet society as they evolved during the course of the Soviet period with Mirovitskaya (1998, p. 32) suggesting three distinct periods when concern for the environment was 'most articulated'. The first of these (1910-1920) reflects the conservation activities of the early Soviet government in addition to the relatively vibrant intellectual legacies of the pre-revolutionary period in the fields of conservation and ecology alluded to above. The second period (1950-1960) acknowledges the emergence of greater civic agency coinciding with the death of Stalin and emerging pollution problems associated with the Soviet Union's industrial activity. The third and final period (late 1980s) is linked with the restructuring policies of the Gorbachev era and the proliferation of grass-roots environmental activity. These distinct periods of environmental expression are useful in providing a framework for the observations of other scholars. For example, the Russian sociologist Oleg Yanitsky (2000, p. 43) has conducted extensive research into the nature of the Soviet Union's 'environmental movement', a term he uses to refer to a distinctive element within Soviet society unified by a concern for conservation and pollution issues. He suggests that the origins of the Soviet Union's environmental movement can be traced back to 1958 thus coinciding with the founding of the first student nature protection organisation at Tartu University in Estonia. A similar organisation was set up in Moscow a couple of years later and then subsequently in a significant number of universities across the Soviet Union. The student organisations, or *druzhiny*, proved particularly adept at maximising their agency within the restrictions of the Soviet regime. For example, whilst often disagreeing with the protestations of the state sponsored All-Union Society for the Protection of Nature (VOOP) they would use their membership rights in order to organise 'citizen inspectorates' related to specific nature protection abuses (Weiner, 1999, pp. 404-405). The environmental stirrings associated with the emergence of a dedicated student movement emerged more obviously during a landmark environmental debate in the Soviet Union surrounding the economic development of Lake Baikal (see Mirovitskaya, 1998, p. 37; Pryde, 1972, pp. 147-151). Proposals to alter the outflow of water from the lake had been made in the late 1950s and these were soon joined by plans to build two cellulose plants on the lake's southern shore. Opposition to the proposed developments was communicated to a wide audience via the influential publication '*Literaturnaya Gazeta*', in addition to other publications, and focussed on scientific shortcomings, bureaucratic inertia and the issue of Russian identity (e.g. Weiner, 1999, pp. 355-373). According to Yanitsky's (2000, p. 44) chronology, the Soviet Union's discernible environmental movement became increasingly visible in the 1980s with the emergence of a range of 'ecologically-oriented' initiatives and the development of urban-based activities, driven to a large extent by general public health concerns. It is important to note the contrast between the practical environmental concerns of the average citizen in the 1980s and the theoretically-informed discussions characteristic of intellectual environmental debate throughout much of the mid-late Soviet period. As Weiner (1999, p. 429) notes, *perestroika* was influential in altering the composition of the environmental movement. Furthermore, the associated policies helped to lay the foundations for the subsequent politicisation of the movement towards the end of the 1980s, with the legalisation of non-governmental activity and the emergence of independent environmental movements within a number of Republics (Yanitsky, 2000, p. 45; see also Weiner, 1999, pp. 429-439 and Chapter 4).

Concluding Remarks

In reviewing the Soviet environmental legacy it is important to highlight a number of issues. First, it is too simplistic to suggest that the Soviet system was characterised by environmental ignorance and an overwhelming contempt for the natural world. This is not to deny that the system gave rise to significant environmental problems, but simply to acknowledge the underlying environmental sensibilities inherent within Soviet society. A preoccupation with the negative aspects of the Soviet Union's environmental legacy, coupled with the regime's marked secretiveness, has tended to deflect attention away from underlying intellectual and political debate, the attempts at administrative and legislative reform and the instances of environmental protest. Second, the notion of environmental crisis formed during the late Soviet period was accommodated more or less uncritically by the liberal democratic consensus, which emerged within both the Soviet Union and the West at this time. As mentioned in Chapter 1, this consensus was predicated on a vision of societal betterment based on the imposition of capitalistic relations and democratic systems of governance. While it is certainly true that significant parts of the former Soviet Union suffered from a range of severe environmental problems, it is also important to acknowledge the vast tracts of land which remained largely untouched by human activity. Third, there has been little attempt to place the experience of the Soviet Union within the context of Western environmental experience beyond obvious comparisons between the West's emerging environmental movement from the 1960s onwards and the Soviet Union's impeded civil society. However, a more general comparison of the environmental characteristics relating to the socialist and Western development models would appear to indicate a degree of similarity. Pryde (1991, pp. 291-292) draws attention to aspects of this overlap by noting the 'striking parallels' between a range of major environmental problems in the Soviet Union and the USA. These include the more obvious Three Mile Island-Chernobyl' comparison as well as instances of environmental pollution related to specific types of industrial activity and nature conservation issues. While such comparisons disguise the underlying causal factors and, just as importantly, the subsequent responses of the state, they do at least point towards the myopia that characterised much of the West's analysis of the Soviet Union's environmental situation during the late 1980s. At a more abstract level, Porritt (1984) in his advancement of a green manifesto, draws attention to the similarities between the capitalist and socialist development models and incorporates both of them under the umbrella term of 'industrialism' in recognition of their joint concern for the continual growth of production output. The nature of the ideological confrontation between the Soviet Union and the West also stimulated the emergence of a range of environmental problems. The localised pollution consequences of the nuclear arms race are an obvious example.

Ultimately, while the Soviet system bequeathed a welter of environmental problems, the Russian Federation also inherited a range of administrative, legislative and intellectual capacities with which to address the evident problems. The tangible and intangible linkages bridging the Soviet and post-Soviet period with respect to issues of environment management are evident in the chapters that follow. Such linkages highlight the fact that while the events of 1991 were certainly significant and represented a fundamental change to the way in which Russian society operated, they did not establish a clean slate on which to graft an entirely new relationship between society and the environment.

Chapter 3

Economic Restructuring and the Wider Environment

Introduction

Chapter 1 outlined three broad approaches towards the Russian environment evident within the general literature since the end of the Soviet period. The first of these revolves around notions of environmental crisis while a second approach focuses on trends of environmental improvement allied to the imposition of capitalistic relations. A third approach operates at a more general level and highlights the significance of the Russian environment for the functioning of global biophysical systems and Russia's role as a global environmental actor. While all three approaches have their limitations they, nevertheless, draw attention to a number of useful trends. Importantly, the different understandings are not necessarily mutually exclusive and various elements are often evident in official documentation, policy reports and academic literature. Taken together they present a rather ambiguous picture of the contemporary Russian environment with images of catastrophe competing alongside those of improvement and global environmental regulation. To a large extent, this ambiguity is related to the various scales (both spatial and temporal) at which the different conceptualisations operate. For example, while the imposition of capitalistic relations has the potential to reduce pollution levels and improve resource-use efficiency within the limited context of a transforming Russian economy, such trends are more problematic when related to the long-term functioning of global ecological systems and prevailing environmental issues such as global warming. At these larger scales of analysis, the environmental deficiencies of the capitalist development model are more easily prioritised as are the long-term global ramifications of Russia's societal transformation. Bearing this in mind, the scope of this chapter is confined to the national and sub-national scale and is concerned with the environmental consequences of Russia's macro-economic transformation over the short- to medium-term and thus focuses on the connections between economic restructuring and environmental trends within the Russian Federation since 1991. Whilst remaining sensitive to the dangers inherent in using Western experience as a means for predicting change within post-socialist countries, it is suggested that the environmental dynamic of Russia's contemporary economic restructuring process can be usefully considered in relation to the attempted construction of a market-type economy. The following section explores general economic and environmental trends before moving on to engage more purposively with the industrial and agricultural sectors of the Russian economy.

Economic Restructuring and Environmental Performance during the 1990s

General Trends

The Russian economy contracted markedly during the 1990s as a consequence of the attempt to construct a functioning market economy on the basis of a well-entrenched command economic system. Indeed, the extent of contraction was so large that it overshadowed the economic dislocation experienced by the USA, UK, France and Germany during the Great Depression years of the early 1930s (World Bank, 2002, pp. 3-5). Between 1990 and Russia's financial crash in August 1998, the level of industrial production more than halved and levels of GDP and agricultural production declined by approximately 45 percent (Goskomstat, 2003a, pp. 179, 202; OECD, 1999, p. 127). During this period of economic decline, the Russian economy also underwent substantial restructuring. The Soviet Union's bias towards military and heavy industrial production was reduced substantially during the 1990s as services increased their proportion of total GDP (Plate 3.1). This restructuring process is reflected in the employment data. For example, while industry remains the main source of employment, it is characterised by a declining trend and accounted for 22 percent of the total workforce in 2002 compared to 30 percent in 1990 (Goskomstat, 1996, p. 84; 2003a, p. 78). Furthermore, as Shaw (1999, p. 120) notes, official statistics underestimate actual levels of service sector development due to the existence of considerable levels of unofficial activity. While it is important to recognise the marked extent of structural economic changes, comparative data highlight the degree to which the contemporary Russian economy differs structurally in relation to advanced capitalist countries (see Table 3.1). Reflecting on structural changes during the course of the 1990s, Gustafson (1999, pp. 219-220) notes the imbalances extant within the Russian economy relative to a market blueprint. In particular, he draws attention to the apparent over-dependence on natural resource export and the boom in trade and service sector activities while certain sectors of manufacturing and agricultural production have suffered greatly giving rise to what he terms a 'barbell' economy. The temptation is, of course, to suggest that Russia's economy will continue to restructure in line with the characteristics of advanced market economies. However, such assumptions require careful revision in the light of possible future trends and regional differentiation.

Table 3.1 Net output (value added) of different sectors as a % of GDP for selected countries, 2001

	Agriculture	Industry	Services
Russian Federation	7	37	56
France	3	26	71
Germany	1	31	68
Japan	1	32	66
UK	1	29	70

Source: World Bank, 2003, pp. 238-239.

The decline in the overall extent of economic activity coupled with the noted structural changes within the Russian economy encouraged a significant contraction in the level of officially recorded pollution emissions. With regard to gross pollution levels, stationary source air emissions reduced by 40 percent between 1991 and 2000 and polluted drainage discharge to surface waters fell by more than one quarter over the same time period. Indeed, declining pollution emissions emerged as one of the defining characteristics of former socialist countries during the 1990s (see Figure 3.1).

(Photograph: author)

Plate 3.1 Symbols of service sector development, central Moscow

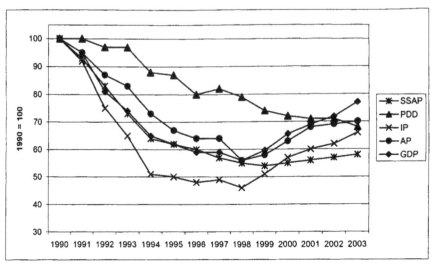

SSAP – Stationary source air pollution
PDD – Polluted drainage discharge
IP – Industrial production
AP – Agricultural production
GDP – Gross domestic product

Sources: calculated from Goskomstat, 1996, p. 252; 2000, p. 19; 2003a, 202; 2003b, p. 21;
2004a, pp. 185, 208; 2004b, pp. 12, 20; OECD, 1999, p. 127.

**Figure 3.1   Comparison of economic output and pollution trends
            (1990 = 100)**

The ten-year anniversary of the Soviet Union's demise prompted a glut of critical
assessments related to the nature of change within the former socialist countries of
CEE and FSU. Analytical reports associated with organisations such as the EBRD
and World Bank highlighted evident weaknesses of performance in relation to the
market blueprint (EBRD, 2001; World Bank, 2002). These centred on issues such
as lack of necessary market discipline, weak economic governance and fragile
financial infrastructures. At the same time, underlying economic inefficiencies and
structural weaknesses were also exposed through the pairing of environmental and
economic indicators, with the suggestion that the Russian economy was
characterised by high levels of resource-use and pollution intensity in comparison
with advanced Western economies and other former socialist countries of CEE.
Such trends were rooted deeply in the Soviet development model and their
persistence reinforced the perception of an inadequate reform process. An OECD
study published in 1999 examined the changing relationship between pollution and
economic performance variables over the course of the 1990s and concluded that
there had been limited decoupling of economic activity from pollution output
(OECD, 1999, pp. 126-127; see also Peterson, 1995a). Crotty's (2002) more

grounded work in Chelyabinsk and Nizhny Novgorod regions is generally supportive of such a trend highlighting the conflict between environmental investment and economic survival at the local level. Crotty concludes that there is '...strong...evidence to suggest economic transition has generated a pollution-intensifying effect' (2002, p. 315). This can be contrasted with trends evident in the former socialist countries of central and eastern Europe where data analysis suggests that a marked decoupling process took place during the 1990s (Zamparutti and Gillespie, 2000).

Energy intensity provides another measure of overall economic efficiency and is calculated based on the energy required to produce one unit of GDP. In the mid-1990s, the Russian economy required nearly 2.5 times more energy in order to produce one unit of GDP compared to the OECD average while, at the same time, levels of energy intensity were actually increasing within certain sectors of the economy (OECD, 1999, pp. 62, 194; see also EBRD, 2001, p. 91). Missfeldt and Villavicenco (2000, p. 382) suggest that falls in Russia's level of CO_2 emissions would have been even greater during the early-mid 1990s had energy efficiency not worsened. The relatively high levels of energy use within the Russian economy can be related in part to the particularities of its climate, a relatively high dependence on energy intensive industries as well as unreported economic activity (EBRD, 2001, pp. 91, 94; UNFCCC, 2000, pp. 3, 11-13, 16-17). At the same time, deep-seated inefficiencies within the Russian economy and, more specifically, the energy production and transmission sector are also highlighted (EBRD, 2001, p. 92; OECD, 1999, p. 193).

Emerging Environmental Problems and Russia's Technical Infrastructure

While gross pollution levels declined markedly during the 1990s, new urban-based environmental pressures emerged associated largely with the aforementioned tertiarisation of the Russian economy. Of particular note were the increased levels of domestic waste production, increased motor vehicle usage and the loss of green space due to urbanising trends in large urban localities such as Moscow (e.g. Bityukova and Argenbright, 2002; Oldfield, 1999; Saiko, 2001). Chapter 5 provides a more detailed overview of changes in the polluting potential of Russia's motor vehicle sector. Comprehensive data related to waste production trends are not available and yet it is clear that waste (both industrial and municipal) represents a growing problem within Russia. The OECD highlights data related to municipal waste generation and these are extrapolated from surveys carried out in a number of Russia's urban regions (Table 3.2). Bearing such caveats in mind, three general conclusions can be reached with some degree of confidence in relation to municipal waste. First, levels of municipal waste generation increased significantly during the 1990s and this can be related to the emergence of a service sector economy, particularly in the country's large urban localities. Second, in spite of these increases, per capita figures are still substantially below those of OECD countries. While future rates of generation may well diverge from those currently found in mature capitalist economies, there would seem grounds to expect a rising trend over the short- to medium-term. Third, Russia's municipal waste management system is heavily biased towards landfill disposal (95 percent in 1992 [OECD, 2002b, p. 14]) and facilities are severely overstretched with limited capacity to recycle effectively.

Table 3.2 Comparative data for municipal waste generation[a]

	Municipal waste (000s tonnes)			Municipal waste (kg/per capita)		
	1980	1990	Late 1990s	1980	1990	Late 1990s
Russian Federation	22000	28000	50000	160	190	340
UK	-	27100	33200	-	470	560
USA	137568	186167	208520	600	740	760
EU 15[b]	125000	148000	188000	370	420	520
OECD average[c]	358000	463000	551000	420	500	560

- Data not available

[a] The original data source advises caution when making cross-country comparisons due to differences in data collection methodologies.

[b] Figure excludes the 10 new accession states.

[c] OECD average includes the following countries: Canada, Mexico, USA, Australia, Japan, South Korea, New Zealand, Austria, Belgium, Czech Republic, Denmark, Finland, France, Germany, Greece, Hungary, Iceland, Ireland, Italy, Luxemburg, Netherlands, Norway, Poland, Portugal, Slovak Republic, Spain, Sweden, Switzerland, Turkey, UK.

Source: compiled from OECD, 2002b, pp. 10-12.

Data for industrial waste production are limited and runs of data are therefore difficult to generate. It has been conventional for Russia's annual State of the Environment Report to focus on trends in levels of hazardous waste generation, although more comprehensive data are now available (e.g. MPR, 2003, p. 200). The aforementioned 1999 OECD report highlighted the fact that levels of hazardous waste generation increased during the 1990s despite the significant falls in levels of industrial production (see also EEA, 2003, p. 155). Furthermore, production levels for this type of waste continued to increase through to 2001 (e.g. Goskomstat, 2003b, p. 49). Taken at face value, this reinforces the notion of an inefficient and polluting industrial system. At the same time, the OECD report acknowledges that this upward trend may be the result of improved reporting procedures (OECD, 1999, p. 90). Suffice to say that industrial waste generation and disposal represents a major concern for Russia, with the 2002 State of the Environment Report highlighting the lack of suitable disposal sites and the tendency for waste to accumulate in unofficial dumps (MPR, 2003, p. 202). Russia also generates significant levels of high-risk waste associated with the country's extensive nuclear and military infrastructure. Once again, it is difficult to determine an accurate picture of the actual levels of this type of waste. The Norwegian environmental group Bellona has focused on aspects of Russia's nuclear waste issue during recent years. For example, in the case of the Russian North Sea Fleet, which comprises an extensive range of nuclear-powered ships, Bellona has highlighted the limited capacity for handling nuclear waste coupled with weak containment infrastructure (e.g. Bøhmer, 1999). Similar concerns have been raised in relation to containment facilities at nuclear reprocessing plants in the southern Urals region (e.g. Bøhmer and Nilsen, 1995).

In addition to emerging environmental issues related to service sector development within Russia's urban localities, more pervasive issues are also apparent. In particular, the deterioration of industrial, military and municipal infrastructure has given cause for concern in Russia and beyond (see Peterson and Bielke, 2002). This trend is related to the depth of structural change within Russian society during recent years, which has undermined the integrity of management and maintenance systems and prevented the implementation of adequate capital replacement programmes. Furthermore, a legacy of marked under-investment stretches back into the Soviet period. Importantly, such concerns are as much applicable to Russia's rural regions as they are to urban localities. At the same time, this should not deflect attention away from ongoing efforts to implement clean-up efforts in certain areas.[1] Some of the noted concern is based on anecdotal evidence such as the recent incidents involving Russia's nuclear submarine fleet or the country's periodic mining accidents. Furthermore, the spectre of another Chernobyl' looms large in the imagination of many Europeans. The Russian nuclear industry has had a relatively high number of incidents during the last decade or so and, while these have been minor for the most part, such trends help to perpetuate perceptions of a flawed nuclear power system (EEA, 2003, pp. 218-219). While remaining sensitive to prevailing Western caricatures of Russian technological systems, available data would appear to provide support for a severely strained industrial, military and municipal technical infrastructure. The State Mining and Industrial Inspectorate (*Gosgortekhnadzor*) provides accident data related to a range of industrial enterprises and affiliated units from key industrial branches such as metallurgy, mining and chemical/petro-chemicals. Recent runs of data suggest a downward trend in the number of accidents with 337 registered in 1995 compared to 207 in 2002, although there was a slight increase between 2002-2003 (Minpriroda, 1996, p. 211; MPR, 2003, p. 202; MChS Rossii, 2004, p. 12). This favourable tendency is related to the more general improvement in the performance of Russia's industrial sector (Peterson and Bielke, 2002, pp. 15-16). Nevertheless, accident rates remain significant and particular concern is centred on the country's main pipeline transport infrastructure. Between 1999 and 2002, accident rates for these objects exceeded 40 counts per annum (see also EEA, 2003, pp. 215-216). The relatively high rates of industrial accidents are linked to organisational and infrastructural weaknesses, financial shortfalls and, in the case of oil pipelines, illegal activities (MPR, 2003, p. 202). Furthermore, Peterson and Bielke (2002, p. 22) underline the fact that Russia's extensive stock of aging industrial infrastructure ensures that large-scale upgrading and replacement will be required in key sectors such as energy production during the

[1] The author would like to thank Nathanial Trumbull for drawing his attention to examples of such activity in St. Petersburg.

course of the next decade or so placing additional pressures on scarce financial resources.

It is perhaps inevitable that any general discussion concerning Russia's economic performance since 1991 dwells on trends of decline and contraction. Nevertheless, more recent developments are worthy of closer examination. Indeed, levels of economic output have started to increase since the low-point of 1998 and there are signs that pollution levels are rising in response to this trend (see Figure 3.1 and Chapter 5). More generally, Russia's industrial and agricultural sectors continue to produce significant levels of pollution at the regional, national and global level despite 8 years of consecutive decline. By 2002, industrial production had recovered to 62 percent of its 1990 level and further growth is expected over the short-term (e.g. Mineconomrazvitiya, 2004). Agricultural production has also recovered in recent years and increased levels of production are envisaged during the course of the next few years. An important element of Russia's relatively strong economic showing since 1999 relates to the export performance of resource intensive sectors of the economy, with mineral products and precious metals accounting for approximately three-quarters of total export value in 2002 or US$70 billion (Table 3.3). Analytical work by Soos et al. (2002, p. 33) concerning Russia's trade relations with the EU (representing Russia's main export market) supports this general observation with resource-intensive (e.g. fuel, non-ferrous metals etc.) and scale-intensive (e.g. heavy chemicals, iron and steel etc.) industries clearly dominating Russia's export profile. The increasing levels of industrial and agricultural output raise the question of how such trends will impact on the wider environment. For example, both Peterson (1995a) and Soos et al. (2002, p. 35) draw attention to the possible deleterious environmental consequences associated with the relative strength of Russia's chemical industry over the short- to medium-term. At the same time, the marked overall fall in pollution output during the last decade has tended to deflect attention away from such concerns. This indifference has been further encouraged by the tendency for medium- and long-term pollution scenarios to utilise the early 1990s as a baseline period for extrapolating future trends. As a consequence, Russia's more recent pollution increases are often submerged beneath the substantial pollution declines of the early to mid 1990s. For example, a recent EEA baseline scenario for predicting air pollution trends in Europe through to 2010 notes that Russia and countries of the former Soviet Union (excluding the Baltic states) will register substantial absolute declines in emission levels across a range of traditional air pollutants over the period 1990-2010 thus downplaying recent upturns in both economic and pollution output (see EEA, 2003, pp. 125-131). The Kyoto Agreement for addressing climate change has been constructed on a similar basis. In order to explore further the trends that lie behind recent gross pollution indicators, the following section focuses on Russia's restructuring industrial sector.

Table 3.3 Russia's foreign trade structure (excluding countries of the Commonwealth of Independent States)

Category	Exports (% of total)			Imports (% of total)		
	1992	1995	2002	1992	1995	2002
Machinery, equipment and transportation	8.9	8.1	8.0	37.7	38.8	38.7
Mineral products[a]	52.1	40.2	56.9	2.7	2.8	1.1
Metals, precious stones etc.	16.4	29.8	20.0	3.3	5.0	5.0
Chemical products, rubber	6.1	9.6	6.4	9.3	11.4	17.8
Timber and cellulose-paper goods	3.7	5.9	4.8	1.2	3.0	4.3
Textiles and textile goods	0.6	1.3	0.6	12.2	4.7	4.4
Foodstuffs & agricultural products	3.9	3.5	1.8	26.0	29.3	24.0

[a] Note this includes fuel-energy resources (*toplivno-energeticheskie resursy*).

Sources: Goskomstat, 1996, pp. 345-346; 2003a, p. 372, 374.

Industrial Restructuring and the Environment[2]

Introduction

As noted above, Russia's industrial sector continues to play a key role with regard to overall pollution discharge levels in spite of the noted falls in output. In particular, it is the main source of stationary source air pollution (SSAP) discharge accounting for over 80 percent of total emissions in 2002 (Figure 3.2). The remaining discharge is attributed largely to the other main sectors of the economy (i.e. transport and municipal sectors). At the same time, the industrial sector's contribution to polluted drainage discharge (PDD) levels is less substantial and accounted for approximately one third of the 2002 total, with the municipal sector responsible for nearly two thirds of overall discharge (Figure 3.3). This represents a significant divergence from polluted drainage discharge trends during the early to mid 1990s when the respective contributions of industry and the municipal sector were roughly equal (Minpriroda, 1996, p. 149).

---

[2] The analysis within this section builds on the work of an earlier paper (see Oldfield, 2000).

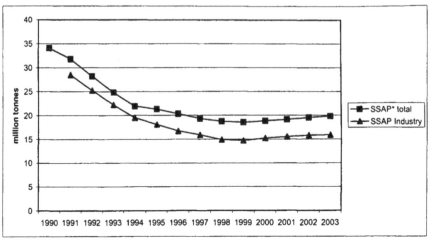

*SSAP – stationary source air pollution

Figure 3.2 Industrial stationary source air pollution emissions

Sources: Goskomstat, 2001a, p. 21; 2004b, pp. 20, 31; Minpriroda, 1996, p. 148a; MPR, 2002, p. 147.

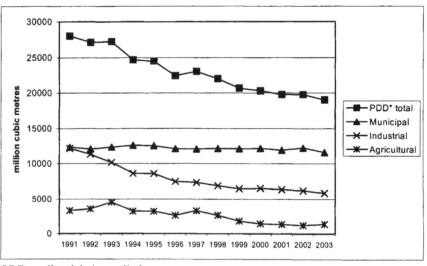

*PDD - polluted drainage discharge

Figure 3.3   Polluted drainage discharge

Sources: Goskomstat, 2001a, p. 21; 2004b, p. 34; Minpriroda, 1996, p. 148a; MPR, 2002, p. 147.

The Soviet system's noted bias towards heavy industry and military output has clearly been undermined in recent years, thus ostensibly removing some of the most pollution-intensive units from production. However, such a generalised statement requires qualification. First, the recent upturn in levels of industrial production is likely, in some instances at least, to draw upon the surplus capacity made redundant during the restructuring of the 1990s. As such, there is the obvious concern that the increased output levels will encourage concomitant excessive increases in pollution emissions. Second, with limited capital available for improving and upgrading infrastructure, there is the likelihood of localised instances of accidental emissions and other inefficiencies leading to raised levels of pollution output. In order to gain a more informed understanding of the polluting potential of Russia's industrial sector in the contemporary period, it is important to move beyond general trends and engage with the particularities of industrial change. An indication of the nature and extent of restructuring can be derived from gross production indicators for individual branches of industry. As highlighted above, both resource- and scale-intensive industrial sectors have tended to respond more favourably to the changing socio-economic situation of the 1990s than the other main industrial branches. As such, electric energy production, fuel and metallurgical (ferrous and non-ferrous) branches of industry recorded below-average declines in levels of industrial production over the period 1990 to 2003 (Goskomstat, 2004a, p. 185). For example, in 1990, these four branches of industry together accounted for approximately 35 percent of Russia's total production output and this had increased to 43.5 percent by 2003 (see Table 3.4). In contrast, the forestry, construction and light industry branches suffered marked levels of production decline over the same time period. As a consequence, their combined contribution to overall production output fell from almost 19 percent in 1990 to approximately 10 percent in 2003. Between these two extremes, chemical/petro-chemical production and engineering branches of industry tended to maintain a reasonably stable share of overall production output relative to other branches.

It should be noted that both resource- and scale-intensive branches of industry tend to be responsible for a disproportionately high level of stationary source air pollution output. For example, in 2002, the electric energy, fuel and metallurgy branches of industry accounted for 85 percent of the industry total and over 70 percent of the Russian Federation total (Figure 3.4). In recognition of their importance to national air pollution levels, the following section explores these industrial branches in more detail.

Russia's Energy Sector

Together with the combined fuel sector, electrical energy production is the major contributor to national levels of SSAP output (approximately 17 percent in 2002; see also Hill, 1999). More specifically, this sector of the economy is the principal source of Russia's overall volume of greenhouse gas (GHG) emissions. The social and economic dislocation of the early 1990s resulted in a substantial fall in the level of electricity consumption and this is reflected in figures for electrical energy

Table 3.4 Structure of industrial production according to main industrial branches, % of total in 1995 prices

Industrial branch	1990	1995	2003[a]
Industry total	100	100	100
Electrical energy	7.9	12.5	9.5
Fuel	12.3	16.6	16.5
Ferrous metallurgy	8.1	9.3	9.9
Non-ferrous metallurgy	6.1	6.6	7.6
Chemical & petro-chemical industry	8.8	8.1	8.4
Machine-building and metalworking	22.9	18.2	19.7
Forestry, woodworking and cellulose-paper	6.1	5.2	4.5
Construction materials	5.7	4.9	3.7
Light	6.9	2.5	1.6
Food	11.8	12.1	13.8

[a] Preliminary data

Sources: Goskomstat, 2000, 35; 2004b, p. 17.

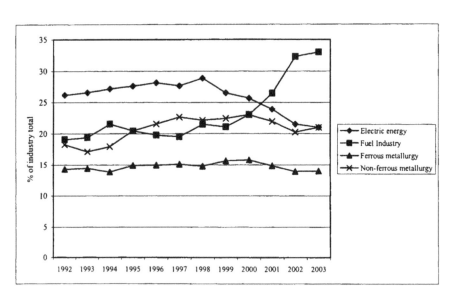

Figure 3.4   Stationary source air pollution emissions for resource- and scale-intensive sectors of the economy, as a % of industry total

Sources: Minpriroda, 1996, p. 148a; Goskomstat, 2004b, p. 31; MPR, 2002, p. 147.

production, which declined by one quarter between 1990 and 1998 (Goskomstat, 2000, p. 148). Unsurprisingly, demand for energy has risen steadily since then in tandem with the recent increases in levels of industrial production (Goskomstat, 2000, p. 148; 2004a, p. 191). Russia's electrical energy production is based on Soviet built infrastructure. Production units are comprised of thermal power stations (TETs – *teploelektrotsentral'*) powered by oil, gas or coal, hydro-electric power stations, and nuclear power stations. At present, thermal power stations are responsible for more than two-thirds of total production output and thus represent the main contributors to national levels of SSAP output. There have been attempts to improve the environmental performance of thermal power stations in recent years focusing, amongst other things, on an effort to substitute oil for gas. Indeed, as noted above, this policy dates back to the Soviet period. Additional remedial activity has included reconstruction work aimed specifically at reducing NO_x and other gaseous emissions. In spite of the noted initiatives, Russia's energy sector remains a significant polluter due in part to persistent inefficiencies. Russia's Energy Strategy, covering the fist two decades of the twenty-first century, provides an indication of the anticipated structure of energy production within the country (Energy Strategy, 2003). Of note is the planned increase in the importance of atomic power within Russia's overall energy production profile, particularly within the context of the strategy's 'optimistic' production scenario (Table 3.5).

Table 3.5 Structure of Russia's electricity production output by source, % of total output

	1990	2000	2020 (moderate production scenario)	2020 (optimistic production scenario)
Thermal powers stations	73.7	66.2	65.1	62.4
Hydro-electric power stations	15.4	18.8	16.0	15.6
Atomic power stations	10.9	15.0	18.9	22.0
Total	100	100	100	100

Source: Calculated from Energy Strategy, 2003, p. 86.

Greenhouse Gas Emissions

Levels of Russia's greenhouse gas emissions merit close attention in recognition of Russia's importance for global trends and the fact that these emissions are associated predominantly with energy production and use. According to data submitted to the secretariat of the United Nations Framework Convention on Climate Change (UNFCCC), Russia was responsible for approximately 17 percent of greenhouse gas (GHG) emissions from 'industrialised' countries (as labelled under the terms of Kyoto) during the base year of 1990 and, as such, was second only to the USA. Russia's Third Communication to the UNFCCC secretariat indicated that emission levels of GHGs had fallen by approximately 40 percent (CO_2-equivalent) between 1990 and 1999 due to the marked fall in levels of industrial activity and associated rates of energy production and consumption (ICPCC, 2002, pp. 8, 37-39; see also Missfeldt and Villavicenco, 2000). The relative fall in the contribution of thermal power stations to overall energy production totals has also been influential in this respect (Table 3.5). According to data published by the OECD (2002a, pp. 43-46), Russia's electricity plants, as well as other facilities involved in energy transformation (e.g. refineries), accounted for approximately 60 percent of overall CO_2 emissions in 1999. In contrast, transport and industry were both responsible for approximately 12 percent of the total. This structure of CO_2 emissions differs significantly from countries such as the UK and USA. In these cases, transport emissions of CO_2 are far more important (24 percent and 30 percent, respectively) whereas energy transformation is correspondingly less significant as a main source (37 percent and 47 percent, respectively). Given the considerable increases in motor vehicle numbers during recent years, it can be expected that the contribution of Russia's transport sector to overall CO_2 emissions will increase over the short- to medium-term.

Fuel and Metallurgical Branches of Industry

According to Russian statistical conventions, the fuel sector of the Russian economy comprises oil extraction, oil refining as well as gas and coal production. In 2002, these industrial branches together accounted for more than 25 percent of Russia's SSAP output. Furthermore, the improved economic performance of this sector since 1998 has encouraged concomitant increases in SSAP emissions within both the oil extraction and coal production branches (Figure 3.4). New accounting methods for calculating pollution discharge levels, in addition to an expanding inventory of industrial units, are also implicated in this increase (MPR, 2003, pp. 152, 162). At the same time, the recent upturn in production levels helps to disguise underlying difficulties. For example, while Russia's oil and gas sectors provided, and continue to provide, substantial support to the domestic economy, levels of investment in exploratory and prospecting work fell markedly during the 1990s. Worsening daily extraction rates reflect the subsequent increased dependence on existing hydrocarbon deposits as mining activities become more complex (Table 3.6). Furthermore, Rotfel'd and Medvedovskii (2003, p. 33) highlight the fact that during the latter half of the 1990s, the growth of proven hydrocarbon reserves failed to keep pace with overall extraction rates. At present, the north western parts of Siberia are the main focal point for Russia's oil and gas industry due to the extensive hydrocarbon reserves located in the autonomous regions of Khanty-Mansi and Yamalo-Nenets. The Yamalo-Nenets region dominates domestic gas production (approaching 90 percent of the total in 2001) whereas oil production is less regionally concentrated with the Khanty-Mansi region accounting for approximately 55 percent of total production output in 2001. Other regions responsible for significant levels of oil production include the Komi Republic (North Western federal *okrug*), Republic of Bashkortostan (Volga federal *okrug*) and the Republic of Tatarstan (Volga federal *okrug*) (Goskomstat, 2002a, pp. 23-24). New sources of hydrocarbon reserves will require development during the next decade in order to ensure proven reserves keep pace with extraction rates and that extraction rates meet those outlined in Russia's aforementioned Energy Strategy. Indeed, this strategy anticipates that by 2020, rates of oil and gas extraction will be approaching, or else exceeding, those recorded in 1990 (see Table 3.7). Eastern parts of Siberia and the Russian Far East are destined to play a much greater role in overall oil and gas production by 2020 with the development of major reserves within these regions. Additional reserves are also known to exist along Russia's continental shelf. Importantly, doubt has been cast upon the ability of Russia to develop successfully these new hydrocarbon reserves due to a combination of inadequate exploratory work, high operational costs and poor infrastructure (see Bradshaw and Bond, 2004; Dienes, 2004). If further development is to take place, this will generate a range of localised socio-environmental tensions as deposits are mined and export infrastructure put in place. Many of these issues are evident, at least to some extent, with respect to the current development of hydrocarbon deposits off-shore of Sakhalin island (e.g. Wilson, 2000; 2002).

Table 3.6 Data related to Russia's oil, gas and coal extraction activity, 2001 as a % 1990

	Oil	Gas	Coal
Annual extraction	67%[a]	91%	68%
Average daily yields from one oil/gas well	66%	74%[b]	-
Extent of prospecting drilling activities (both oil and gas)	35%	35%	-

[a] Includes condensates
[b] According to the data of RAO Gazprom

Source: calculated from Rotfel'd and Medvedovskii, 2003, pp. 33-43.

Table 3.7 Actual and predicted extraction rates for oil, gas and coal resources, 1990-2020

	1990	2000	2020 (moderate production scenario)	2020 (optimistic production scenario)
Oil (mln tonnes)	516	324	450	520
Gas (bln m^3)	640	584	680	730
Coal (mln tonnes)	395	258	375	430

Source: Energy Strategy, 2003, pp. 62, 72, 81.

Russia's coal production is concentrated predominantly in five main regions and these include the Kuznetsk basin (Kemerovo *oblast'*), Donets basin (south-eastern European Russia),[3] Kansk-Achinsk basin (Krasnoyarsk *krai*), Pechora basin (Komi Republic) and the southern Irkutsk region (MPR, 2003, p. 162). Russia's coal mining regions are amongst the country's most polluted according to measures of air and water quality and this is related strongly to the industry's relatively poor technical state (Rotfel'd and Medvedovskii, 2003, pp. 42-43). Importantly, according to the aforementioned Energy Strategy, the coal industry is to remain an integral part of Russia's energy base during the medium term with increased output envisaged over the course of the next 15 years (see Table 3.7) and with particular emphasis placed on the Kuznetsk and Kansk-Achinsk coal basins (Energy Strategy, 2003, pp. 82). This obviously has implications for the overall polluting potential of Russia's energy sector as well as environmental trends at the local level.

Russia's metallurgical industry (ferrous and non-ferrous) was responsible for 28 percent of Russia's stationary source air pollution in 2002. Economic activity related to ferrous metallurgy (iron and steel production) is focussed predominantly in south central Siberia, southern European Russia (iron ore, steel smelting) and the southern Urals region (pig iron, steel smelting, rolled ferrous metal and steel tube production) (MPR, 2003, p. 166). The non-ferrous sector (i.e. aluminium, copper, nickel, zinc production etc.) incorporates the infamous Noril'skii Nikel' combine, which is situated above the Arctic circle in eastern Siberia (see Hønneland, pp. 33-36). In 2002, the urban locality of Noril'sk was responsible for more than 2 million tonnes of SSAP, representing 10 percent of Russia's total discharge of stationary source pollution (Goskomstat, 2003b, pp. 65-66).

Agricultural Restructuring and the Environment

Up until now relatively little has been said concerning Russia's agricultural sector and this reflects the tendency within the general literature to focus attention predominantly on urban-based economic trends. This imbalance makes little sense given the significance of agricultural activity within the Russian economy and Russian society more generally (Unwin et al., 2004). For example, the agricultural sector comprised approximately 11 percent of the total workforce in 2003 and 5.4 percent of GDP (Goskomstat, 2004a, p. 79; 2004b, p. 12). Furthermore, land utilised for agricultural purposes across all land categories (*sel'skokhozyaistvennye ugod'ya* or 'farmland') accounted for approximately 13 percent of Russia's total land area in 2002 (MPR, 2003, p. 32). It should be noted that Russia's most productive agricultural land is concentrated in the southern parts of European Russia (Chernozem or 'Black Earth' region; see also Chapter 5).

Russia's agricultural sector was characterised by significant structural change in the aftermath of the Soviet Union's demise (e.g. Ioffe and Nefedova,

---

[3] Note that the main part of this coal basin is located in Ukraine.

2004; Pallot and Nefedova, 2003; Unwin et al., 2004). In the early 1990s, approximately three-quarters of Russia's agricultural output was attributed to agricultural enterprises, in essence the successors to collective and state farms. Ioffe and Nefedova (2001, p. 390) draw attention to this historical link by referring to such entities as 'socialised farms'. The importance of these types of farms was predicted to recede during the 1990s as private forms of farming emerged. However, this expected trend failed to emerge with private farms accounting for only a small percentage of overall agricultural output by 2000 and the remnants of state and collective farms continuing to provide a framework for emerging agricultural production networks (e.g. Ioffe and Nefedova, 2001; Pallot and Nefedova, 2003). Interestingly, under Putin the emphasis has been placed more firmly on large-scale farming as the way forward for Russia's agricultural sector (Wegren, 2002, p. 28). At the same time, personal/household subsidiary farming (production attributable to the private plots of individual citizens) emerged as the main source of agricultural production, accounting for over 50 percent of agricultural output in 2002 compared to approximately 25 percent in 1990 (Goskomstat, 1996, p. 550; 2003a, p. 201). Pallot and Nefedova (2003, pp. 41-42) caution that such a trend is not simply indicative of rising self-sufficiency amongst the population in the face of social dislocation. Indeed, their *rayon*-level work in several regions of European Russia highlights the complex nature of this activity consisting, as it does, of varying degrees of self-sufficiency as well as commercial activity. It should also be noted that the distinction between socialised farms and subsidiary farming is not necessarily as clear-cut as the statistics suggest with close connections evident between the two entities. For example, Ioffe and Nefedova (2001, pp. 399-400) posit that some of the recent increase in subsidiary farming output may actually be a side-effect of socialised farms re-classifying parts of their own output in order to avoid tax payments.

The marked contraction in the extent of Russia's agricultural activity during the 1990s suggests that this sector's polluting potential and propensity to disrupt local ecosystems should have declined accordingly. Following the OECD approach, this potential can be inferred from a range of proxy data such as the level of mechanisation, energy-use and chemical application rates within the sector. Data specific to the Russian Federation are certainly indicative of a significant decline in the agricultural sector's polluting potential during the course of the 1990s (see Figure 3.5). More particularly, levels of polluted drainage discharge attributable to agricultural activity fell by approximately 60 percent between 1991 and 2001, a trend connected closely with the marked falls in livestock numbers over the given period (Goskomstat, 1996, p. 570; 2001b, p. 72; see also Figure 3.3). At the same time, there is evidence to suggest that the structural changes within the agricultural sector during the early-mid 1990s have had a detrimental impact on land and soil management systems in certain regions (e.g. MEPNR, 1994, p. 41; 1995, p. 16; MPR, 2002 p. 35-36; Peterson, 1995a, p. 305). According to official reports, soil erosion affected at least one-quarter of Russia's agricultural land in the early 1990s (MEPNR, 1994, p. 41) and the area of arable land subject to erosion is reported to increase annually by approximately half a million hectares (e.g. MPR, 2003, p. 34). Different types of land degradation tend to predominate in particular regions. For example, wind and water erosion is most acute in the agricultural regions west of the Urals. In contrast, salinised soils are most apparent in parts of Siberia and the Volga region (IOPRR, 2001, pp. 107-108). The process of desertification is also notable in a large number of Russia's southern regions stretching from the North Caucasus to eastern parts of Siberia (MPR, 2003, p. 34). Chapter 5 provides a more detailed exploration of changes in the area of farmland during the course of the 1990s.

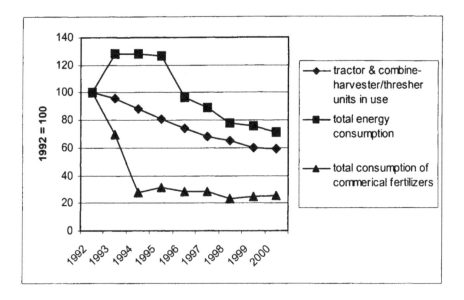

Figure 3.5   Trends in Russia's agricultural sector, 1992 = 100

Source: Calculated from OECD, 2002c, pp. 11-19.

Regional Patterns of Economic Change and Associated Environmental Trends

The general economic trends outlined above can disguise marked variation in the nature of economic change across the Russian Federation and related environmental trends. Some scholars have focused attention on the contrast between the relative economic success of a handful of urban centres and the difficulties being faced (both economic and social) in the intervening spaces. For example, Dienes (2002, p. 444) draws comparisons between the frontier hinterlands of northern European Russia, Siberia and the Russian Far East and what he terms Russia's 'inner' hinterland (*glubinka*) located mainly in European Russia and the Urals region. He posits that this inner hinterland is characterised by:

> ...large tracts of virtually dead space that interpenetrate and separate the economically responsive archipelago of the integrated economy...The primitive hinterlands between and beyond the major metropolitan centers remain deprived of the most elementary

physical and social infrastructure. In contrast to Western Europe, but also the whole eastern half of the United States and its Pacific states, Russia's urban nuclei have been able to draw into their social-economic web only parts of the interstices between metropolitan areas.

Dienes' conceptualisation of contemporary Russian socio-economic space more than hints at the existence of marked regional inequalities and this interpretation is supported by the variation in the performance of different economic sectors coupled with their associated geographies of production during the course of the 1990s. Regional difference with respect to socio-economic characteristics is not restricted solely to the post-Soviet period. Indeed, while the Soviet development model was grounded on notions of spatial equality, in actual fact socio-economic variation was often marked. With the imposition of capitalistic relations, however, patterns of uneven development have been further reworked (see Bradshaw and Vartapetov, 2002) and thus reflect trends witnessed more generally across CEE (e.g. Hamilton, 1999).

The immense size of the Russian Federation undermines effective regionalising frameworks and throws the tensions evident between national-level generalisations and local-level complexities into stark relief. At the same time, a feel for regional patterns of economic change can provide a useful framework within which to approach Russia's contemporary environmental situation. A number of political and economic studies carried out during the mid to late 1990s attempted to highlight emerging regional trends (e.g. DeBardeleben and Galkin, 1997; see also the review by Shaw, 1999, pp. 94-126). With regard to economic regionalisations, the work of Bradshaw and Hanson (1998; see also Hanson and Bradshaw, 2000a) was particularly effective at drawing attention to emerging patterns. Reflecting on the existing literature as well as their own detailed case study work, Bradshaw and Hanson (1998, p. 289) noted that the '[A]nalysis of regional economic performance reveals a complex interplay of inherited advantages and constraints, which combine with the ability of the regional élite to orchestrate economic recovery.' Their resulting understanding of Russia's emerging regional economic geography was thus based on a combination of factors including: political leadership, industrial infrastructure, natural resource endowment, level of agricultural activity and the extent of new types of economic activity associated with the service and financial sectors. In addition, urban localities displaying relatively high levels of connectivity in relation to international and/or national economic processes also performed comparatively well during the 1990s, according to a range of primary economic indicators. These included Moscow city, Samara (middle Volga) and Ekaterinburg (Urals). Such regions tended to attract significant flows of FDI (Bradshaw, 2002), although as indicated in Chapter 1, overall levels of FDI for Russia are considered low in the international context. While a reasonably small number of regions were able to take advantage of the changing economic circumstances post-1991, large expanses of Russia suffered considerably during the 1990s. This was particularly evident in those regions characterised by a heritage of heavy industry and manufacturing activity in addition to many rural regions.

Table 3.8 A regionalisation of environmental trends in the Russian Federation during the 1990s

Regional types based on economic criteria	Key socio-economic trends during the 1990s	Key prevailing environmental pressures	Indicative urban & federal regions exemplifying this category type
Large urban localities acting as a focal point for regional, national, or international activities	relatively rapid levels of marketisation; construction of office and retail space; relatively significant growth in the size and extent of the service sector; relatively significant levels of FDI[a]; influx of western goods	increasing levels of municipal waste production; increasing levels of motor vehicle pollution; increasing pressure placed upon green areas due to office, retail and housing construction	e.g. Moscow city, St. Petersburg, Nizhny Novgorod, Samara, Novosibirsk, Vladivostok
Regions dominated by heavy industry, engineering, manufacturing and light industry	marked declines in levels of industrial production during the 1990s, recent signs of recovery in some sectors (e.g. petro-chemicals); social pressure to keep large enterprises operational; service sector development	falling levels of gross industrial pollution during the 1990s; increasing pollution intensity in some industrial sectors; relatively weak technical/environmental infrastructure exacerbating environmental pressures	e.g. Vologda, Novgorod, Belgorod, Lipetsk, Sverdlovsk, Chelyabinsk, Kemerovo, Samara
Regions characterised by substantial reserves of natural resources – forestry, mineral extraction, hydrocarbon extraction	reorientation of production output to foreign markets (both Western and Pacific Rim); relatively high levels of FDI*; key role in Russian exports; relative neglect of processing activities	inefficient extraction techniques; extensive localised pollution particularly in the case of hydrocarbon extraction; relatively high levels of corruption and illegal activity due to weak management systems	e.g. Republic of Karelia, Komi Republic, Republic of Sakha (Yakutia), Krasnoyarsk *krai*, Taymyr (Dolgano-Nenets) AO[b], Evenki AO, Tyumen', Khanty-Mansi AO, Yamalo-Nenets AO, Sakhalin
Rural regions	marked restructuring of agricultural production; dislocation of management systems; significant fall in levels of production and land maintenance activities in some regions	falling levels of chemical use; increasing levels of erosion and deterioration of soil fertility and quality; deteriorating infrastructure leading to water pollution, chemical spills etc.	e.g. Republic of Kalmykia, Adygeya Republic, Republic of Dagestan, Krasnodar *krai*, Stavropol *krai*, Altay *krai*, Republic of Tuva

[a] FDI - Foreign Direct Investment   [b] Autonomous *okrug*

Regional understandings of economic change invite connections to be made with the wider environment based on admittedly rather rudimentary linkages between economic activity and polluting trends. Furthermore, while acknowledging that regional environmental particularities are the consequence of a myriad of economic, social, political and biophysical factors operating at a range of scales, a generalised typology relating economic activity to environmental concerns is nevertheless useful in encouraging a differentiated understanding of Russia's regional environmental situation. During the mid 1990s, Peterson (1995a) identified six regional blocs based on broadly defined economic activities in order to distinguish environmental pressures and speculate on future trends. In addition to the obvious need to relate different environmental trends to regions characterised by relatively high levels of industrial and agricultural activity, he also distinguished the oil and gas producing region of north-west Siberia and the urban regions of Moscow and St. Petersburg as localities with distinctive environmental characteristics.

Table 3.8 extends Peterson's work as well as internalising elements of the work of Bradshaw and Hanson (1998) referred to above in order to establish a generalised framework for approaching regional environmental issues within Russia during the first decade of reform. The categories are not intended to be predictive, but simply provide an indication of the actual and potential variation in environmental problems evident across the Russian Federation in the contemporary period. Furthermore, the high-level of generalisation ensures that the categories are not necessarily mutually exclusive and certain regions will exhibit trends characteristic of more than one category.

Concluding Remarks

The scope of this chapter has been limited to the particularities of the relationship between economic restructuring and measurable environmental trends during the course of the 1990s. The marked fall in gross pollution emissions allied to a downsizing industrial base and restructuring economy has received considerable attention in the Western literature. At the same time, it is worth noting a number of important trends related to this general observation. First, Russia's industrial base remains a significant polluter within the context of both the domestic economy and also the global economic system. The particularities of Russia's restructuring process, with the tendency to fall back on natural resource extraction and export activities, have ensured that the more pollution-intensive branches of industry, such as energy production and metallurgy, remain influential within the national economy. At the same time, new environmental pressures have emerged associated with the burgeoning service sector economy and these include motor vehicle pollution and waste production. Wider, structural issues should also be noted. Russian inherited a techno-infrastructure in need of considerable investment and the relatively limited availability of investment capital during the 1990s has simply aggravated an already strained situation. The polluting potential of Russia's restructuring agricultural sector has also been transformed over the course of the last decade or so. Gross pollution output has fallen with declines in production. Nevertheless, there is evidence to suggest that land management practices are sufficiently weak in certain regions to exacerbate the loss of soil through erosion processes. The final part of the chapter attempted to advance the work of Peterson (1995a) in order to posit a generalised framework of regional environmental issues across the Russian Federation. While this framework is not intended to be predictive, it does encourage the development of a more differentiated understanding of Russia's contemporary environmental situation. Chapter 5 provides more insight into the nature of Russia's regional environmental trends since the early 1990s.

Chapter 4

Governing the Environment: The Place of the Environment in Russia's Political-Administrative System

Introduction

The central aim of this chapter is to determine the place of environmental concern within Russia's political-administrative system post-1991. First, Russia's developing environmental legislation and policy base is assessed through a focus on official documents and associated literatures. Second, Russia's emerging environmental administrative framework is examined with a particular emphasis placed on the changing nature of environmental protection infrastructure. A third section explores the relationship between government structures and non-governmental activities. The concluding section provides an overview of the changes in these three areas and highlights ongoing issues and problems.

It is difficult to avoid conceptualising 1991 as a marked watershed for Russia's environmental policy, with the emphasis on underlying trends of betterment and improvement. However, such an approach would overlook the debates active during the late Soviet period concerning environmental policy (see Chapter 2). Furthermore, the range of both intangible and tangible continuities between the Soviet and post-Soviet periods should not be underestimated. For example, a number of articles concerning the environment elaborated in the 1977 Soviet Constitution were to find their way, albeit in modified form, into the new Russian Constitution of 1993. In addition, Russia's main environmental law, which was promulgated in 1991 and underpinned related legislative developments throughout the 1990s, was conceived during the final years of the Soviet Union. The roots of Russia's ecological expertise,[1] environmental monitoring network and protected area system also lie very firmly in the Soviet period.

Russian societal trends during the early 1990s tended to mirror, at least in a general sense, those characterising the *perestroika* years under Gorbachev. Both periods were marked by attempts to restructure society and accompanied by critical

---

[1] According to the 2002 environmental protection law (article 33) '[T]he ecological expertise is to be carried out with the aim of establishing the conformity of planned economic and other activity to environmental protection demands'. The expertise is supposed to be an independent procedure, although recent organisational changes have caused some commentators to question its autonomy.

reflections on the relationship between Russian society and the wider environment. It is therefore of little surprise to find some semblance of similarity in the aims of environmental policy during these two periods. For example, the abridged English-language version of the 1988 USSR Report on the State of the Environment indicated that the '[E]cologization of the economy is a key trend of perestroika' (Goskompriroda, 1990, p. 35) and this has considerable overlap with the environmental rhetoric surrounding the purposeful movement to the market in the 1990s. The essence of these continuities has been reshaped and reformed during subsequent years via the influence of changing political priorities and by prevailing international trends, encapsulated by the main tenets of sustainable development. It is also worth dwelling on the upsurge of environmental concern, which characterised the last years of the Soviet Union and the extent to which this spilled over into the post-Soviet period. Such concern was underpinned by a combination of growing nationalist fervour within a number of Union Republics as well as the strengthening agency of civic movements vis-à-vis the state. Galvanised by the latent pool of talent within the Soviet Union's environmental movement, and the relative freedom of expression associated with environmental issues, wide sections of Soviet society gathered under the environmental banner in order to vent their frustrations at the state and its central bodies of power (e.g. Peterson, 1993; Ziegler, 1992). As such, environmental concern became intimately wrapped up with the fall of the Soviet Union (see Tickle and Welsh, 1998) and environmental improvement became an almost given condition of post-socialist change within many quarters. This enthusiasm for, and belief in, environmental betterment characterised debate in the West as well as much of CEE and the FSU during the late 1980s/early 1990s, and was a key factor behind the 'environmental euphoria' which swept the region at this time (e.g. see Pavlinek and Pickles, 2000, pp. 171-172; and also Fagin, 2001; Tickle and Welsh, 1998). Leading 'green' figures at the local level were absorbed by Russia's fledgling democratic system (e.g. Yanitsky, 2000, p. 45), although it is likely that certain political actors utilised the environmental issue in order to advance their own political positions during the early 1990s (Van Buren, 1995, p. 130). The late Soviet period thus provides an important background to the subsequent changes in environmental legislation, policy and related administrative structures that took place across the FSU. In the case of Russia, the environment remained an influential political theme during the first years of the 1990s. This was aided greatly by the establishment of the RSFSR Ministry for Ecology and Natural Resources in late 1991 on the basis of former republic and All-Union structures, thus raising the prominence of the environment within the country's decision-making process. However, such influence was destined not to last. Political co-option and opportunism were important factors contributing to the gradual erosion of the environmental movement's political influence during subsequent years. Furthermore, a significant cause of the growing environmental apathy evident within Russian society more generally as the 1990s progressed was the widespread deterioration in the quality of life.

The following section begins with an examination of Russian environmental policy[2] and legislation development since 1991 in order to provide an indication of its underlying nature, emerging logic and direction.

Russian Environmental Legislation and Policy Initiatives

The Late Soviet Period

Mention was made in Chapter 2 of the Soviet Union's reasonably extensive legislative base in the area of nature protection and the rational use of natural resources. Furthermore, this general concern has a long history traceable to the early Soviet period and, indeed, the pre-revolutionary era. During the Soviet Union's final two decades, the authorities became increasingly aware of the need to address environmental problems in a complex rather than a piecemeal manner and this was highlighted by the aforementioned 1972 Supreme Soviet decree 'Concerning Measures for the Further Improvement of Nature Protection and the Rational Utilisation of Natural Resources' (Kostenchuk et al., 1993, section 4.2.1). The impetus for such an initiative, while certainly related to developments within Soviet environmental policy formation, cannot be wholly divorced from activities at the international level. The United Nations Conference on the Human Environment took place in Stockholm during the same year and brought together 113 states with the intention of discussing '...the need for a common outlook and for common principles to inspire and guide the peoples of the world in the preservation and enhancement of the human environment' (Stockholm Declaration, preamble) (see also Caldwell, 1984). While the Soviet Union was not a participant at the meeting, the general outcomes of the conference stimulated discussion and debate within the country's policy and academic communities.

The noted 1972 decree appeared to herald the emergence of a distinct phase in the development of Soviet environmental legislation characterised by a purposeful 'ecologisation' (*ekologizatsiya*) of the legislative base and the gradual development or, in some cases, initiation of bodies of law for areas such as water (1970), minerals (1975), forests (1977) and air quality (1980) (see also Lisitzin, 1987). The growing importance of the environment within Soviet policy was further reinforced by the inclusion of relevant articles within the 1977 Soviet Constitution (see Table 4.1). The early 1970s was also the period in which a more coherent system of environmental data collection emerged (Dumnov, 2002, p. 41). Legislative development in the area of natural resource use and environmental protection continued during the 1980s and was related to the working through of internationally agreed environmental commitments, the elaboration of general environmental laws at the regional level (e.g. in relation to the preservation of Lake

---

[2] It should be noted that Russian official documentation refers to 'ecological' policy (*politika v oblasti ekologii*). However, in recognition of the general scope and nature of this policy in the Russian context, the term environmental policy is used as a suitable synonym throughout this chapter.

Baikal) and the continued 'greening' of the Soviet Union's law base at a more general level (Kostenchuk et al., 1993, section 4.2.1).

Environmental Legislation and Policy Initiatives Post-1991

The reasonable extent of Soviet environmental legislation complicated greatly the process of legislative revision and renewal within Russia post-1991. Furthermore, in common with other branches of law, the 1990s were characterised by a period of intensive law creation specifically in the area of nature protection and natural resource use (Table 4.2). In trying to make sense of the underlying rationale of legislative developments in this area, it is necessary to explore in more detail the main laws and policy documents promulgated during this period. The 1993 Russian Constitution makes various allusions to environmental issues and concerns thus providing a potentially firm basis from which to ground an effective environmental legislative and policy framework. The most relevant articles are outlined in Table 4.1 and there is some continuity with elements of the 1977 USSR Constitution. However, one of the most influential legislative developments in the area of environmental protection predated the 1993 Constitution. The aforementioned 1991 RSFSR law 'Concerning the Protection of the Natural Environment' was issued just a few days before the dissolution of the Soviet Union and thus provides an insight into Soviet thinking concerning the relationship between society and nature during the late Soviet period (see Bond and Sagers, 1992).[3] Indeed, it was not revised in full until 2002. The opening paragraph of the 1991 law gives an indication of its underlying logic stating that:

> Nature and its riches are the national property of the Russian people, the natural basis for their sustainable social-economic development and the well-being of humankind.

Article 1 goes on to outline the tasks of nature protection legislation in the Russian Federation as:

> ...the regulation of the interrelationship between society and nature with the aim of preserving natural wealth and the natural environment of humankind, preventing the ecologically harmful influence of economic or other activities, enhancing and improving environmental quality and strengthening justice and law and order in the interests of current and future generations.

These statements are interesting as they capture, in part at least, the general concerns of the international community at that time, with their allusion to notions of sustainable socio-economic development and both intra- and intergenerational equity. Such sentiments were to find powerful backing at the UN Conference on Environment and Development (UNCED), held in Rio de Janeiro the following year, which endeavoured to advance the notion of sustainable development at the

---

[3] See also OECD (1996) for contextual information related to Russia's 1991 environmental law.

Table 4.1 Environmental coverage in the 1977 USSR Constitution and the 1993 Constitution of the Russian Federation

Fourth Constitution of the USSR (1977)	Constitution of the Russian Federation (1993)
Section: The Economic System	Section: Fundamentals of the Constitutional System
Article 18 – 'In the interests of the present and future generations, the necessary steps are taken in the USSR to protect nature and to make scientific, rational use of land and its mineral and water resources, and the plant and animal kingdoms, to preserve the purity of air and water, ensure reproduction of natural wealth, and improve the human environment'	*Article 9 [1]* – 'The land and other natural resources are to be utilised and protected in the Russian Federation as the basis of the life and activity of the peoples inhabiting the corresponding territory'
Section: The Basic Rights, Freedoms and Duties of Citizens of the USSR	Section: The Rights and Liberties of Citizens
Article 42 – 'Citizens of the USSR have the right to health protection. This right is ensured by free, qualified medical care provided by state health institutions; by extension of the network of therapeutic and health institutions; by the development and improvement of safety and hygiene in industry; by carrying out broad prophylactic measures; by measures to improve the environment; by special care of the health of the younger generation, including prohibition of child labour, excluding the work done by children as part of the school curriculum; and by developing research to prevent and reduce the incidence of disease and ensure citizens a long and active life.	*Article 42* – 'Everyone has the right to a favourable environment, reliable information about the state of the environment and compensation for damage caused to health or property by ecological offences'
Section: The Basic Rights, Freedoms, and Duties of Citizens of the USSR	Section: The Rights and Liberties of Citizens
Article 67 – 'Citizens of the USSR are obliged to protect nature and conserve its riches'	*Article 58* – 'Everyone is obliged to protect nature and the environment and to show solicitude for natural wealth'

Sources: Constitution, 1985; Konstitutsiya, 1997; also see Pryde, 1991, pp. 6-7.

global level. Importantly, Russia was a participant at this conference and made a positive contribution by signing up to the various instruments and declarations and these included: Agenda 21, the Rio Convention as well as the Convention on Biodiversity and UN Framework Convention on Climate Change (UNFCCC). Agenda 21 is particularly important as it lies at the heart of the global drive towards sustainable development and represents a 'comprehensive programme of action' which '...reflects a global consensus and political commitment at the highest level on development and environment cooperation' (Article 1.2, Preamble). Signatories to Agenda 21 are expected to develop a national strategy in order to engage with the multiple facets of the developmental and environmental issues raised by the initiative. The preamble to Agenda 21 acknowledges the differing capacities of the signatories to address the demands of the programmatic action and calls for international cooperation in order to mitigate some of the evident inequalities. Moreover, it notes the acute nature of the social, economic and political difficulties faced by Russia and other so-called transition economies. Due to the wide-ranging nature of Agenda 21, the underlying concept of sustainable development has played a role in the evolution of Russian environmental legislation and policy during the 1990s within the context of much broader social and economic concerns.

Table 4.2 Selected list of key environmental laws (*zakony*) and codes (*kodeksi*) promulgated since 1991

Name of Law	Year
Concerning the Protection of the Natural Environment	1991[a]
Concerning the Ecological Expertise	1995[a]
Concerning Specially Protected Natural Areas	1995
Concerning Wildlife	1995
Water Code	1995[b]
Forest Code	1997[c]
Concerning Wastes of Production and Consumption	1998
Concerning Protection of the Atmosphere	1999
Concerning the Sanitary-Epidemiological Well-Being of the Population	1999
Concerning the Protection of the Environment	2002[d]

[a] Note that additions and changes were made to these laws during subsequent years.
[b] New edition under consideration.
[c] A new version of the Forest Code is currently under review.
[d] This law superceded the 1991 law Concerning the Protection of the Natural Environment.

Source: compiled by author; see also OCED, 1999, p. 46.

Russian Interpretations of Sustainable Development

Sustainable development is a wide-ranging concept with its roots in Anglo-US understandings of the relationship between society and nature. The history of the concept's emergence to claim its current influential position in global environmental policy formation has been well-rehearsed in the general literature, and is intimately connected with broader developments in international environmental debate and cooperation extending back to the aforementioned Stockholm Conference in 1972 (e.g. see Dryzek, 1997; Macnaghten and Urry, 1998, pp. 212-219). The Brundtland Commission's oft-cited elaboration of the concept's meaning provided the foundation for the 1992 Rio conference (WCED, 1987).[4] Generally speaking, sustainable development is concerned with ensuring the conditions for continuous economic growth based on capitalistic relations and allied to improving social and environmental conditions. Furthermore, the concept is grounded on notions of intra- and inter-generational equity. Sustainable development is thus not restricted to environmental concerns, but has pretensions to emerge as a main unifying policy agenda at the global level bringing together hitherto largely disconnected debates and policy-making in areas such as poverty reduction, social equality and justice. The concept has been attacked for its apparent perpetuation of the global *status quo* via its underlying support of existing economic and political power structures (e.g. Sachs, 1999). As such, it is argued that the concept side-steps the need for serious debate related to the radical transformation of social and economic systems in order to reconcile them with ecological systems (e.g. Foster, 2002, pp. 79-82). At the same time, the concept is characterised by certain positive characteristics such as its ability to bring a diverse range of actors together around the same table in order to debate key global issues. In addition, notions of sustainable development, as well as related ideas of sustainability, have encouraged the development of numerous practical and applied initiatives aimed at mitigating humankind's negative impact on the surrounding environment at various stages of the capitalist production cycle.

The loftier core concerns of sustainable development, with their aim to ensure a resolution of the conflict between humankind and the wider environment, provide scope for finding common ground with pre-existing indigenous sensibilities and aspirations. Russia's engagement with the concept of sustainable development would certainly appear to reflect this tendency (see Oldfield, 2001; Oldfield and Shaw, 2002). However, while the concept is having some influence guiding the evolution of Russia's medium- and long-term socio-economic and environmental policy, it has provoked notable debate within the country's scientific literature over its meaning and relevance in the Russian context. One area of debate concerns the inadequacy of the Russian

---

[4] According to the Brundtland Commission, '[S]ustainable development is development that meets the needs of the present without compromising the ability of future generations to meet their own needs' (WCED, 1987, p. 43).

translation of sustainable development as '*ustoichivoe razvitie*'. This translation fails to capture the essence of sustainable development and is more closely related to notions of 'stable development', encouraging some Russian commentators to formulate alternative definitions in order to provide greater conceptual clarity (see Oldfield and Shaw, forthcoming). More generally, elements of the critical debate in Russia are characterised by similar concerns to those found in the Western literature, and relate to the underlying efficacy of the concept. At the same time, other authors are largely uncritical in their acceptance of the concept and, drawing on qualities purportedly embedded in Russian society, focus on Russia's apparent empathy with the central tenets of sustainable development articulated within a general framework of society-nature balance.[5] Russia's relatively advanced scientific and technical capabilities are also advanced as a basis for the successful implementation of sustainable development at both the national and global scale. The promotion of Russia's importance for the state of the global environment forms a key aspect of the domestic political rhetoric concerned with sustainable development and associated environmental policy (see Oldfield et al., 2003).

The Incorporation of the Sustainable Development Concept into Russia's Legislative and Policy Base

As indicated above, Russia was an active participant at the Rio 1992 conference and the concept of sustainable development has influenced a range of policy and legislative developments during the course of the last decade or so. In February 1994, a Presidential decree was passed 'Concerning the Russian State Strategy for Environmental Protection and Ensuring of Sustainable Development'.[6] The preamble to this decree acknowledged the influence of the 1992 UN conference in Rio. In addition to outlining the main elements of state policy in the given area, the decree was also designed to form the basis of an environmental action plan covering the period 1994-1995.[7] This action plan was to incorporate four main areas:

- Ensuring of ecologically secure sustainable development within the framework of a market economy;
- The protection of human environments;
- The rejuvenation (restoration) of damaged ecosystems in ecologically unsafe regions;
- Russia's participation in the solution of inter-governmental and global ecological problems.

---

[5] See Rozenburg et al., 1996 for a general discussion.
[6] For a copy of the decree see Rossiiskaya gazeta, 9 February, 1994, p. 4.
[7] See Sobranie zakonodatel'stva Rossiiskoi Federatsii, 1994, Vol. 4, 23 March, pp. 590-608.

The 1994 decree did not represent a detailed elucidation of Russia's official understanding of sustainable development. Instead, it remained at a general level and connected sustainable development with notions of 'ecologising' the country's industrial production system and reducing pollution levels. Furthermore, it glossed over the noted difficulties of translating the concept into the Russian language.

Discussions concerning the nature of the concept were carried out at an official level during the period 1994-1996. A further Presidential decree in 1996 outlined Russia's 'Concept for the Transition of the Russian Federation to Sustainable Development'. This was a direct response to the demands of Agenda 21, which, as indicated above, required each participating country to outline its own strategy for sustainable development.[8] This connection is made clear in the opening paragraph of the decree:

> Following the recommendations and principles laid down in the documents of the UN conference for environment and development (Rio-de-Janeiro, 1992), and acting in accordance with them, it is considered necessary and possible to realise the successive transition to sustainable development in the Russian Federation, ensuring the balanced solution of social-economic issues and the problems of preserving a favourable environment and natural-resource potential, in order to satisfy the needs of current and future generations (Ustoichivoe razvitie, 1996, p. 6).

The 1996 decree begins by outlining the necessity of achieving sustainable development at a global level before indicating the specifics of the Russian situation. Key policy areas are highlighted and these include the establishment of a supportive legislative base as well as necessary economic mechanisms. In addition, emphasis is placed on the need to ascertain the ability of environmental systems to cope with human interference, at both the local and regional level, as well as the need to disseminate the idea of sustainable development widely amongst the general public. Brief sections cover regional issues, indicators for determining sustainable development and the importance of international cooperation. The decree ends by outlining a three-staged transition to sustainable development for Russia (see Oldfield [2001] for a more detailed critique). In the short-term this involves attaining a balance between socio-economic activity and the environment and the fulfilment of ameliorative activities. The second (mid-term) stage is to involve substantial structural changes in order to facilitate a 'greening' of the Russian economy. The final stage moves to the global level and envisages 'the harmonious interaction of the world community with nature' and suggests that such a state would coincide with '...the emergence of the noosphere (sphere of reason) as envisaged by the eminent Russian scientist V.I. Vernadsky when the spiritual values and knowledge of humankind, living in harmony with the environment, become the measure of national and individual wealth'. Space precludes a detailed discussion of his ideas, yet it is important to offer some additional insight here.

---

[8] For a copy of the 1996 Presidential decree see: Rossiiskaya gazeta, 9 April, 1996, p. 5.

Vladimir Ivanovich Vernadsky (1863-1945) was a Russian scholar of considerable ability who engaged in a wide variety of scholarly pursuits in addition to carrying out a range of civic duties (see Bailes, 1990). Vernadsky's conceptualisation of the noosphere was grounded in his work on the biosphere, which he developed over a significant part of his academic career. The fundamentals of his understanding concerning the structure and nature of the biosphere emerged from an interest in the active connections between organic and inorganic matter and were elaborated in his book by the same title (Vernadsky, 1998). This field of enquiry can be traced back to the work of V.V. Dokuchaev as well as other Russian intellectuals active during the mid and late nineteenth century (Oldfield and Shaw, 2002; forthcoming). For Vernadsky the biosphere was envisaged as a coherent whole characterised by an immense potential to influence the geology of the Earth's surface via the functioning of living matter. A fundamental aspect of his conceptualisation was the belief in the progressive expansion of the area of the biosphere over time. Furthermore, he gave humankind an increasingly prominent role, suggesting that we are emerging as *the* main geological force within the biosphere, and he related this characteristic to the development of our collective intellectual capabilities made manifest in scientific thought. He thus envisaged the transformation of the biosphere into the noosphere as intimately associated with humankind's mental capacities, and compared it with other major periods of the biosphere's evolutionary history such as the emergence of large expanses of forest millions of years ago (see Vernadsky, 1945). While the noosphere concept is proving influential as a means for conceptualising the end-point of sustainable development within Russia, there would appear to be a marked disjuncture between the largely symbolic use of the noosphere idea within the 1996 decree and Vernadsky's attempt to develop a convincing scientific framework underpinning its emergence (see Oldfield and Shaw, in press, for a more detailed analysis of Vernadsky's ideas).

Russia's state strategy for sustainable development remains in draft form despite a number of revisions. Furthermore, the extent to which the 1996 decree and associated initiatives have influenced wider policy developments is clearly open to question and debate. Nevertheless, the discourse surrounding Russia's understanding of sustainable development provides an insight into the way in which society-nature relations are being conceptualised more generally by the Russian government. For example, besides the above reference to Vernadsky and the noosphere, parts of the 1996 decree draw attention to the perceived close affinity between Russia's 'customs, spirit and mentality' and the core concerns of sustainable development, thus mirroring similar statements in the Russian social science literature. Furthermore, the connections being made between something intrinsically Russian and notions of society-environment balance are maintained in subsequent policy statements, and provide a means for buttressing Russia's claim that it has a key role to play in ensuring sustainable development at the global level (see Oldfield and Shaw, 2002; Oldfield et al., 2003). Ostensibly, the documentation referred to above is largely the product of state, academic and, to a lesser extent, non-governmental organisation interaction. However, it is unclear how effectively the different groups are incorporated into the policy-making process. Russia's indigenous peoples represent an additional interest group with much to offer the ongoing debate concerning the nature of sustainable development. Recent work highlights the different ways in which such groups relate to nature, or else are being affected by the consequences of marked societal transformation (e.g. see Crate, 2002; Wilson, 2003). As such, their understanding of what constitutes a 'sustainable' way of life is likely to differ from the vision expounded in the 1996 decree and related policy pronouncements.

Russia's Environmental Legislation and Policy Base from the Mid-1990s Onwards

The discourse surrounding sustainable development prompted some rethinking of legislative and policy direction in the areas of socio-economic development and environmental protection within Russia, and this is reflected in associated documentation and legislative advances. While the overall effectiveness of such advances is at best limited, elements of key environmental legislation and associated policy initiatives established during the 1990s are either reducible to the central tenets of what might be termed conventional sustainable development, or else are proactive in their advancement of the concept. In other words, these documents reflect a general concern for improving the efficiency of natural resource use and reducing the pollution-intensity of the country's industrial system with efforts to connect such 'ecologising' intentions to wider social issues (e.g. see Oldfield, 2001). Concerns including the preservation of biodiversity levels and the expansion of protected land areas are also evident (see Table 4.2). More specifically, legislation and policy initiatives related to strategic concerns, such as national security and socio-economic development, are suggestive of deliberate efforts to internalise elements of sustainable development associated with key areas of society-environment interaction. The formulation of a Medium-Term Programme for the Socio-Economic Development of the Russian Federation covering the period 2002-2004 is interesting due to its advancement of an 'ecologically-oriented' economy '...characterised by minimal negative influence on the environment and high resource and energy effectiveness' (Programma, 2001, p. 116) as well as its influential policy recommendations. For example, it promoted an improvement in the efficiency of nature protection administrative organs, the formation of a complex programme of action in order to provide a focus for state environmental policy and the revision of Russia's basic environmental law. It also restated Russia's importance in tackling global environmental problems. In line with these suggestions, a federal 'target' programme, 'Ecology and Natural Resources of Russia (2002-2010)', was approved by the Russian government in early December 2001. Target programmes are a key policy tool providing the means for implementing state policy by channelling finance to clearly delineated activities. This particular programme is an attempt to rationalise the plethora of programmatic environmental activities that were advanced within Russia during the 1990s. The effectiveness of these multiple programmes had been undermined by inadequate finance and weak coordination

(Ekologiya, 2001, pp. 13-14). The fundamental aim of the programme 'Ecology and Natural Resources of Russia (2002-2010)' is the:

> ...balanced development of the natural resource base to meet the needs of the economy for fuel-energy, mineral, water, forest and biological water resources and to ensure the constitutional rights of citizens to a favourable natural environment' (Ekologiya, 2001, p. 5).

The first stage of the programme (2002-2004) was concerned predominantly with addressing pollution issues in ecologically unfavourable regions of Russia and strengthening the country's natural resource base. Stage two (2005-2010) is more proactive and focuses on reducing pollution levels and ensuring a sustainable utilisation of natural resources over the medium- and long-term. These two stages have overlap with the more general staged transition to sustainable development proposed by the 1996 Presidential decree 'Concerning the Concept of Russia's Transition to Sustainable Development'. The main programme consists of 12 sub-programmes in order to ensure the effective implementation of the overall aim, and these are listed in Table 4.3.

Table 4.3   Sub-programmes comprising the Federal target programme 'Ecology and natural resources of Russia, 2002-2010'

Sub-programmes

Mineral raw material resources

Forest

Water resources and water objects

Aquatic biological resources and aquaculture

Regulation of environmental quality

Wastes

Support for specially protected natural territory

Conservation of rare and endangered fauna and flora

Protection of Lake Baikal and the natural territory of the region

Restoration of the river Volga

Hydrometeorological ensuring of vital 'life' activity (*zhiznedeyatel'nost'*) and the rational use of natural resources[a]

Progressive technologies for cartographic-geodesy provision

[a] This concerns the attempt to conserve the technical base of hydrometeorological services and environmental monitoring networks within the context of market restructuring.

Source: compiled from Ekologiya (2001).

As indicated above, the Medium-Term Socio-Economic Development Plan (2002-2004) highlighted the need for a revision of the 1991 environmental protection law in order to address perceived weaknesses in nature protection activity and ecological standardisation and regulation (Programma, 2001, p. 118). This prefaced the promulgation of a new law 'Concerning the Protection of the Environment' in January 2002. It has already been suggested that the 1991 law received favourable reviews in recognition of its ability to guide the development of Russia's environmental legislation base during the initial transition period. However, the fundamental reworking of federal-regional relations during the 1990s, combined with deep-seated societal changes, ensured that elements of the 1991 law had become ineffectual (Vasil'eva, 2002, p. 1). Superficially, the new law has a similar structure to that of its predecessor (see Table 4.4). However, whereas the 1991 law was conceptualised as having a key role to play in guiding related legislative developments, the 2002 law is conceived as a more complex entity forming '...not only a code of normative demands, but also a jurisdictional interpretation of the foundations of the state's environmental policy' (Vasil'eva, 2002, p. 4). The preamble to the 2002 law expresses sentiments similar to those found within the early sections of the 1991 law (see above), although there is specific reference to the 1993 Constitution and the concept of sustainable development:

> In accordance with the Constitution of the Russian Federation each person has the right to a favourable environment, each person is obliged to preserve nature and the environment, to be mindful of natural wealth which is the basis for sustainable development, life and the activities of the peoples living on the territory of the Russian Federation. The current federal law determines the legal basis of state policy in the area of environmental protection, ensuring the balanced solution of socio-economic tasks, the preservation of a favourable environment, biological diversity and natural resources with the purpose of satisfying the demands of current and future generations, and the strengthening of environmental law and order and the ensuring of ecological security.

Nevertheless, in spite of the grand designs for the new law, it has not been received positively by all sides, with some commentators suggesting it fails to build on the legislative advancements made during the early to mid-1990s (see Larin et al., 2003, pp. 295-296).

Table 4.4 Comparison of the main sections found within the 1991 and 2002 laws for environmental protection[a]

Main chapters of the 1991 law for environmental protection	Main chapters of the 2002 law for environmental protection
I. General principles	I. General principles
II. The rights of citizens to a healthy and favourable environment	II. The fundamentals of environmental protection administration
III. The economic mechanism of environmental protection	III. The rights and duties of citizens and public and other non-commercial associations in the field of environmental protection
IV. Setting environmental quality norms	IV. Economic regulation in the field of environmental protection
V. State ecological expertise	V. Setting norms in the field of environmental regulation
VI. Ecological requirements for the location, planning, construction, reconstruction and initiation of enterprises, construction projects and other activities	VI. Evaluation of environmental influence and the environmental expertise
VII. Ecological demands for the operation of enterprises, facilities and other objects	VII. Environmental protection requirements related to the realisation of economic and other activity
VIII. Extreme ecological situations	VIII. Ecological disaster zones, ecological emergency situation zones
IX. Specially protected natural areas and associated features	IX. Natural objects having special protection
X. Environmental monitoring and control	X. State environmental monitoring
XI. Ecological education and research	XI. Control in the field of environmental protection
XII. Resolution of disputes concerning environmental protection	XII. Scientific research in the field of environmental protection
XIII. Liabilities for environmental offences	XIII. Fundamentals for forming an ecological culture
XIV. Compensation for damages caused by the violation of environmental law	XIV. Liabilities for breaching environmental protection legislation and resolution of disputes in the field of environmental protection
XV. International cooperation in environmental affairs	XV. International cooperation in the field of environmental protection
	XVI. Concluding points

[a] See Bond and Sagers (1992) for a comprehensive overview of the 1991 law.

World Summit on Sustainable Development 2002: Russian Responses

In the autumn of 2002, a 10-year review of the Rio process was carried out under the sponsorship of the United Nations in Johannesburg, South Africa. The World Summit for Sustainable Development (WSSD) was characterised by a preparatory period in which the specifics of Russia's sustainable development policy were reviewed domestically. Two initiatives are worthy of additional comment. First, the Ministry for Economic Development and Trade was primarily responsible for putting together an assessment of Russia's progress towards sustainable development in preparation for the Johannesburg summit (Mineconomrazvitiya, 2002). The choice of this particular ministry to spearhead Russia's response is suggestive of the relative importance placed upon the effective elaboration of a sustainable development policy by the Russian state, at least at the level of rhetoric. In addition, it also reflects the key significance placed on economic development for the realisation of sustainable development over the long-term. This document was characterised by a broad approach to sustainable development and engaged with economic, social and ecological issues from both a domestic and international perspective. The parameters against which Russia's progress was measured remained consistent with previous documentation and underlined the importance of establishing effective democratic and market-type infrastructure in addition to preserving natural systems and addressing issues of environmental quality (Mineconomrazvitiya, 2002, Introduction). Section 5.12 of the document highlighted a number of perceived threats related to Russia's achievement of sustainable development within the ecological/environmental sphere. These included the production of industrial and domestic waste, the country's ageing industrial techno-structure (with particular reference to the hydrocarbon developments in the far north), trans-border pollution issues and climate change. Once again, much of this covered familiar ground and dovetailed with existing concerns. Finally, section 6 drew attention to the need for greater coordination between the different initiatives and programmes related to the implementation of sustainable development. It also reinforced Russia's considerable potential for realising sustainable development, noting the country's extensive scientific capacity, natural resource endowment as well as existing technical capabilities. Its potential prominence as a world leader in the global transition towards sustainable development was also noted.

The second development requiring further comment relates to President Putin's request for the elaboration of an 'Ecological Doctrine' in order to underpin state environmental policy. In addition, this doctrine was to be presented at the World Summit on Sustainable Development in Johannesburg in 2002. Both non-governmental organisations (NGOs) and academics were able to exercise influence over the content of the doctrine, although there was some evidence of these groups being marginalised during the latter stages of its development. The final version of the doctrine was approved by government decree on August 31, 2002, with the stated aim of determining state environmental policy over the long-term (IOPRR, 2002b, p. 120). The doctrine opens with reference to its constitutional underpinnings while also acknowledging the influence of the 1992 Rio conference and subsequent international forums related to issues of the environment and sustainable development. According to the doctrine, the underlying aim of state environmental policy is:

> ...the conservation of natural systems, supporting their integrity and life-support functions for the sustainable development of society, raising the quality of life, improving the health of the population and the demographic situation and ensuring the ecological security of the country (IOPRR, 2002b, p. 120).

The doctrine also outlines the main principles on which state environmental policy is grounded (IOPRR, 2002b, p. 120). These acknowledge the need for a broad engagement with the sustainable development concept in addition to a reduction in the deleterious environmental consequences of economic activity and the active involvement of key social groups in addressing environmental issues. The full list of key priorities is as follows:

- Sustainable development - paying equal attention to its economic, social and environmental elements and recognising the impossibility of societal development in conjunction with environmental degradation;
- High importance of the biosphere's life-support functions with respect to the utilisation of resources;
- Fair distribution of income from the utilisation of natural resources and access to such resources;
- Prevention of negative environmental consequences resulting from economic activity and an assessment of the widespread environmental effect of such activity;
- Prohibition of economic and other projects having unpredictable consequences for the environment;
- Payment for resource utilisation and compensation for damage inflicted on the population and the environment as a result of the infringement of environmental laws;
- Openness of ecological information.

The means for carrying out state policy effectively are discussed and include improvements to Russia's environmental management and monitoring systems, the development of economic and financial mechanisms for regulating the use of natural resources and pollution output, international cooperation and support for relevant areas of scientific activity and education. Sections are also devoted to the importance of ecological education, non-governmental activity and regional ecological policy.

After more than a decade of change and development, Russia's environmental policy, in its broadest sense, would appear to possess a number of discernable characteristics (see also Programma, 2001, p. 116; MPR, 2003, p. 341). These are focussed on concerns to reduce the negative environmental consequences of economic activity through modernisation policies and to establish effective regulatory mechanisms, ranging from complex environmental assessment procedures to fining and payment systems. Emphasis is also being placed on the active involvement of a wide range of actors in order to address environmental issues in a sophisticated manner. Russia's generally positive engagement with global environmental issues is a further recurring feature of recent policy documents. Nevertheless, and echoing the situation during the Soviet period, it is clear that there remains a marked gap between rhetoric and the concrete implementation of the stated policies. Furthermore, some critics go so far as to suggest Russia's contemporary environmental policy is in a state of progressive degeneration, lacking scientific rigour and clear direction from above (e.g. Larin et al., 2003; Mazurov, 2004). The reasons for the evident weaknesses are wide-ranging and related strongly to the marked nature of societal change during the course of the last decade. At the same time, there are continuities with deficiencies in policy implementation and environmental governance apparent during the Soviet period. These are particularly evident in the case of Russia's evolving administrative infrastructure concerned with environmental protection.

The Evolution of Russia's Environmental Management Infrastructure

The Soviet Legacy

Environmental protection emerged as a key area of concern for government activity within many Western countries during the 1970s, and this opened up the political space necessary for the centralisation of monitoring and regulatory functions. The creation of the US Environmental Protection Agency is perhaps the most obvious example (see EPA, 1992). In contrast, and as noted in Chapter 2, while attempts were made to integrate environmental concerns more purposefully within central planning bodies, the Soviet Union maintained a fragmented approach to the environment, with control and management functions dispersed amongst a variety of ministries and state committees. Legislative restructuring in the early 1970s reinforced the primary leadership role of the Supreme Soviet and the Council of Ministers in the area of environmental protection and natural resource use, while comprehensive planning and management functions were devolved to a dedicated section of *Gosplan* (Kostenchuk et al., 1993, section 4.1.1). At the same time, management of natural resources was largely the responsibility of the appropriate economic ministry or state committee.

The weakness of the Soviet Union's administrative system with regard to controlling natural resource exploitation and ensuring environmental protection was considered during the mid- to late-Soviet period (e.g. Pryde, 1972, pp. 13-24; Taga, 1976). Criticism tended to focus on the underlying structure of economic incentives, the aforementioned fragmented nature of the system and the absence of a strong, independent regulatory body capable of challenging the powerful economic ministries. Towards the end of the Soviet period, attempts were made to address some of the evident shortfalls of the environmental protection system culminating in the creation of a State Committee for the Environment (*Goskompriroda SSSR*) in 1988. *Goskompriroda SSSR* was intended to facilitate the realisation of a 'unified scientific-technical policy' in the area of environmental protection and, perhaps more importantly, to coordinate the activities of the various government bodies involved in nature protection (Kostenchuk et al., 1993, section 4.1.1). This was no simple task as it required the development and cultivation of effective relations with a host of entities ranging from the Supreme Soviet to the branch level management structures of ministries and departments. In addition, *Goskompriroda SSSR* was not the only administrative body with environmental protection functions operating at the inter-branch level. By 1991, other bodies included the State Hydrometerological Service (*Gosgidromet SSSR*), the State Committee for Forests (*Goskomles*), the State Atomic Energy Inspectorate (*Gospromatomenergonadzor*) and the Ministry for Health (*Minzdrav*) (Kostenchuk et al., 1993, section 4.1.2). Unfortunately, the inherent conservatism of the Soviet administrative system ensured that existing ministries and departments resisted *Goskompriroda*'s interference in areas of competence traditionally considered part of their own remit (DeBardeleben, 1990, pp. 251-252; Peterson, 1993, pp. 162-168). Nevertheless, the nature of the Soviet Union's final years, entwined as they were with heightened environmental awareness incorporating large sections of society, resulted in *Goskompriroda* being upgraded to a Ministry at the All-Union level during 1991. A similar administrative restructuring took place within the RSFSR, and the ensuing ministerial designation was retained in the administrative framework of the Russian Federation. In spite of this and related developments, the long and contested process of establishing an independent environmental body within the Soviet Union's administrative system failed to leave a lasting legacy in the case of Russia. During the course of the 1990s, environmental concern was destined to lose out to the more powerful claims of the natural resource use lobby. This can be blamed on a combination of the noted inherited structural weaknesses as well as prevailing conditions within Russian society. The Russian Federation therefore acquired a relatively extensive administrative system concerned with nature protection, and yet one undermined by internal disharmony.

Administering the Russian Environment during the Early-Mid-1990s

The ministerial designation given to environmental protection functions ensured that they played a relatively prominent role in Russian policy formulation during the first part of the 1990s. The creation of the RSFSR Ministry for Ecology and Natural Resources in late 1991 represented a marked centralising trend with environmental protection functions combined with regulatory responsibilities. Nevertheless, uncertainty over the scope of the ministry's competencies was reflected in a change of name (Ministry for the Protection of the Environment and Natural Resources) and a reconstituted internal structure during the early 1990s. Importantly, this name change coincided with a number of former natural resource committees reasserting their independence, reflecting the continuation of the entrenched departmentalism of the Soviet period. Furthermore, behind the name-change, a more fundamental shift in state attitudes towards environmental governance was apparently taking shape, and this surfaced in 1996 with the administrative downgrading of environmental protection functions and their subsequent transferral to the newly formed State Committee for Environmental Protection (*Goskomekologiya*). The loss of ministerial representation reduced the influence of environmental protection concerns at the highest levels of government policy-making. According to its founding decree, *Goskomekologiya* was '…to realise government policy in the area of environmental protection, to ensure ecological security and the preservation of biological diversity and to bear responsibility for the improvement of environmental quality' (Ukaz, 1997b). Its overall areas of competence could therefore be reduced to control and monitoring functions, the introduction of pollution and related standards and environmental assessment procedures. *Goskomekologiya* also retained an extensive territorial infrastructure with regional committees in the majority of federal units (OCED, 1999, Pp. 48-50). During the same administrative reshuffle, the Ministry for Natural Resources was formed out of the State Committee for Geology and Mineral Use (*Roskomnedr*) and the State Committee for Water (*Roskomvoda*). In hindsight, this was to mark the beginning of the ascendancy of the natural resource lobby within the government administrative infrastructure. The founding decree also endowed the new Ministry for Natural Resources with environmental protection functions stating that it was:

> …a federal organ of executive power, to carry out state policy in the area of the study, renewal, utilisation and protection of natural resources (Ukaz, 1997a).

The remit of the Ministry therefore had obvious overlaps with that of *Goskomekologiya*, and the two administrative organs were required to coordinate their activities in certain areas.

Environmental Management under Putin

Since Putin's accession to power in 2000, Russia's environmental protection infrastructure has undergone further structural change. This has been closely associated with the far-reaching reorganisation of the state administrative infrastructure underpinned by Putin's determination to reassert central control over Russian territory. In May 2000, Putin issued a decree 'Concerning the Structure of Federal Organs of Executive Power' in which he abolished both *Goskomekologiya* and the Federal Forestry Service. The competencies of these two organs were transferred to the aforementioned Ministry for Natural Resources. This action aroused considerable opposition within both Russian and Western environmental communities (Peterson and Bielke, 2001, pp. 69-70). In particular, there was great concern over the possible consequences of transferring environmental protection functions to a Ministry with the rationale to develop and utilise the country's resource base. This concern was most acute in relation to the functioning of the State Ecological Expertise and its ability to function independently. During the course of the following two years, four State Services were established within the Ministry for Natural Resources concerned with environmental protection, forestry, minerals and water resources. This administrative structure at least ensured that environmental protection functions maintained a relatively coherent position within the context of the ministry. The ministry also consolidated its presence at both the regional level and within the infrastructure of the newly formed federal *okruga*. Despite the apparent radical nature of administrative change at this time, Crotty's work in Samara *oblast'* (Crotty, 2003) suggests that it had limited immediate impact at the local level, with regional environmental protection activity functioning under the Ministry for Natural Resources much as it did under *Goskomekologiya*. Nevertheless, it is important to remain sensitive to the particularities of her case study region and, as Crotty notes, to acknowledge the possibility that the experience of other federal regions with significant natural resource endowment differed substantially.

A further shake-up of Russia's executive administrative infrastructure occurred in March 2004 with the promulgation of another Presidential decree (see Ukaz, 2004). Once again this resulted in significant changes being made to the shape of Russia's environmental protection administrative framework. In particular, the decree resulted in the establishment of one Federal Service and three Federal Agencies placed under the jurisdiction of the Ministry for Natural Resources and these coincided, more or less, with the aforementioned State Services. These new entities included (see also Figure 4.1):

- Federal Service for Inspection in the Sphere of Ecology and Nature-Use (which would later become simply the Federal Service for Inspection in the Sphere of Nature-Use);
- Federal Agency for Forestry;
- Federal Agency for Water Resources;
- Federal Agency for the Utilisation of Mineral Resources.

According to the March 2004 decree, Federal Services (*federal'nye sluzhby*) are empowered with control and supervisory functions within their given area of competence whereas Federal Agencies (*federal'nye agentstva*) are endowed with management functions related to the ascribed area of competence as well as law-enforcement functions (Ukaz, 2004, articles 4 and 5). In the context of the newly reformed state administration, the Federal Service for Hydrometeorology and Environmental Monitoring (*Rosgidromet*) was confirmed as a federal organ of executive power in June 2004.[9] Additional state organs responsible for environmental monitoring activities include the Ministry for Health (sanitary-epidemiological assessments), Ministry for Civil Defence, Extraordinary Situations and Prevention of Natural Disaster (*MChS*) and the Federal Service for Ecological, Technological and Atomic Inspection. The constant restructuring of Russia's nature protection organs, combined with the existence of a range of representative governmental bodies and organisations, ensures that it is very difficult to determine the overall effectiveness of nature protection activity.

---

[9] See Government decree No. 372, 23/06/2004.

In the light of recent changes, the current Ministry for Natural Resources represents a powerful executive body responsible for a range of environmental protection, control and supervisory functions, which had until 2000 been the jurisdiction of independent bodies. The wide remit of the ministry is reflected in Government decree No. 370 (22 July, 2004), which states that:

> The Ministry for Natural Resources of the Russian Federation (MPR Russia) is a federal organ of executive power, realising functions for the elaboration of state policy and normative-legal regulation in the area of the study, utilisation, renewal and protection of natural resources…' (Article, 1, Section 1).

It then proceeds to list a range of policy areas and these include:

- management of the state mineral fund and forest economy;
- utilisation and protection of both the water and forest funds;
- exploitation and maintenance of reservoirs and the water provision system;
- utilisation of wildlife and related habitat (except hunting resources);
- specially protected territories;
- environmental protection (with the exception of the ecological expertise/inspection).[10]

The newly constituted ministry would appear to have a number of pressing issues to contend with. First and most obviously, it must try and convince a largely sceptical group of interested observers (both in Russia and the West) that it is capable of facilitating the effective operation of environmental control and supervisory functions in the face of the understandable urge to develop the natural resources under its jurisdiction. Second, it must ensure the development of a coherent and focussed environmental policy grounded on reliable and realistic financing. Third, it must generate positive collaboration with non-governmental organisations, business groups and other actors within Russian society in order to engender a more open discussion of natural resource development and related environmental issues.

---

[10] Note that this final point was added to the original text by a later decree, which established the working parameters of the Federal Service for Inspection in the Sphere of Nature-Use. See Figure 4.1.

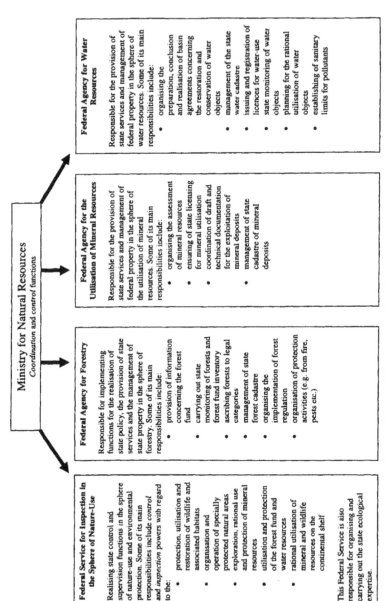

Ministry for Natural Resources
Coordination and control functions

Federal Service for Inspection in the Sphere of Nature-Use

Realising state control and supervision functions in the sphere of nature-use and environmental protection. Some of its main responsibilities include *control* and *inspection* powers with regard to the:

- protection, utilisation and restoration of wildlife and associated habitats
- organisation and operation of specially protected natural areas
- exploration, rational use and protection of mineral resources
- utilisation and protection of the forest fund and water resources
- rational utilisation of mineral and wildlife resources on the continental shelf

This Federal Service is also responsible for organising and carrying out the state ecological expertise.

Federal Agency for Forestry

Responsible for implementing functions for the realisation of state policy, the provision of state services and the management of state property in the sphere of forestry. Some of its main responsibilities include:

- provision of information concerning the forest fund
- carrying out state monitoring of forests and forest fund inventory
- ascribing forests to legal categories
- management of state forest cadastre
- organising the implementation of forest regulation
- organisation of protection activities (e.g. from fire, pests etc.)

Federal Agency for the Utilisation of Mineral Resources

Responsible for the provision of state services and management of federal property in the sphere of the utilisation of mineral resources. Some of its main responsibilities include:

- organising the assessment of mineral resources
- ensuring of state licensing for mineral utilisation
- coordination of draft and technical documentation for the exploitation of mineral deposits
- management of state cadastre of mineral deposits

Federal Agency for Water Resources

Responsible for the provision of state services and management of federal property in the sphere of water resources. Some of its main responsibilities include:

- organising the preparation, conclusion and realisation of basin agreements concerning the restoration and conservation of water objects
- management of the state water cadastre
- issuing and registration of licences for water-use
- state monitoring of water objects
- planning for the rational utilisation of water objects
- establishing of sanitary limits for pollutants

Sources: Government decrees: Nos. 282, 283 (16 June, 2004), No. 293 (17 June, 2004), No. 400 (30 July, 2004). For copies of the various decrees in Russian see the Ministry for Natural Resources website: www.mpr.ru (accessed August 2004) or the website of Rossiiskaya gazeta: www.rg.ru/ (accessed September 2004).

Figure 4.1 Federal organs of executive power under the jurisdiction of the Ministry for Natural Resources

Russia's Environmental Movement and its Engagement with State Structures

Contemporary policy documentation demonstrates an apparent concern amongst state organs to engage with civic action groups and movements. For example, the effective implementation of sustainable development is typically grounded on a purposeful and productive relationship between state and non-state actors. At a more general level, the expected burgeoning of what is often termed 'civil society' within Russia and other former socialist states represented a main theme running through the transition literature during the early to mid 1990s. Since this time, many authors have questioned both the nature and meaning of such a term within the context of post-socialist societal change (e.g. Pavlinek and Pickles, 2000, pp. 164-169). In particular, the largely uncritical (not to mention imprecise) understanding of civil society forwarded within certain sections of the literature, whereby Western experience is used as a key indicator of expected trends, reinforces the linear conceptualisation of change discussed in Chapter 1. This type of approach can encourage the use of proxy indicators (e.g. the number of non-governmental organisations) in order to judge 'progress', and yet these are often incapable of exposing the complex nature of less formal civic action evident throughout the post-socialist region. Furthermore, a reliance on such indicators can result in alternative environmental protest strategies being branded 'deviant' and ineligible for foreign financial and technical assistance. Fagan and Jehlička (2003) provide an interesting critique of conventional attempts to theorise the nature of environmental movements within post-socialist countries using case studies from the Czech Republic. In particular, they note the common failure to internalise the complexity of Western environmental movement activity within prevailing analytical frameworks. In other words, programmatic approaches to post-socialist environmental movements are often grounded on a simplified understanding of Western experience. Fagan and Jehlička also draw attention to the importance of contextual factors (e.g. economic, political) operating at a range of scales (e.g. domestic, EU), which ensure that the development of Czech environmental movements is typically more complex than indicated by the oft-noted trends of increasing institutionalisation and professionalism.

It was suggested in Chapter 2 that, in a general sense, environmental sensibilities have an extended lineage in Russian society. In addition, it was pointed out that according to the Russian sociologist, Oleg Yanitsky, the contemporary Russian environmental movement has its roots in the Soviet period and, more particularly, the student nature protection movements, which appeared in the 1950s and 1960s (Yanitsky, 2000, pp. 1, 43; see also Weiner, 1988; 1999). The historical depth and continuity associated with varying degrees of environmental activism evident within Russia during both the Soviet and pre-revolutionary period reinforce the need to recognise the latent capacity for such action embedded in contemporary Russian society. At the same time, it is just as important to pay heed to the way in which changing political and economic circumstances influence the nature of environmental action. For example, as intimated in the introduction to this chapter, environmental issues provided a convenient focal point for a range of social grievances at the end of the 1980s and beginning of the 1990s, which served to distort their overall level of importance. In addition, environmental concern also provided aspiring politicians of all leanings with useful political capital during the democratic elections of the nascent Russian state post-1991. The societal changes caused by the fall of the Soviet Union precipitated a marked reorganisation of Russia's environmental movement as it fractured along both ideological and structural lines during the course of the 1990s (see Smith, 1999; Yanitsky, 2000). In common with observations elsewhere in CEE and the FSU (e.g. Jancar-Webster, 1998), a key aspect of this dynamic period for Russia was the emergence and subsequent consolidation of professionalised environmental groups characterised by Western-style organisational structures and an ability to access foreign funding initiatives. In Russia's case, professionalised groups include country branches of international organisations such as Greenpeace and the World Wide Fund for Nature as well as domestic initiatives (e.g. Socio-Ecological Union). The aforementioned work of Fagan and Jehlička (2003) encourages caution in arriving at definitive understandings of this phenomenon, with the suggestion that professionalised groups may well have sacrificed domestic environmental concerns for Western-sponsored global agendas. At the same time, other types of environmental groups are also evident. These range from movements centred on a local issue (e.g. Garb and Komarova, 1999, pp. 165-191) to educational groups and societies such as the All-Russian Society for the Protection of Nature (VOOP), which has its origins in the early Soviet period (see Weiner, 1999).

Relations between Russia's environmental movement and state structures have been characterised by periods of marked conflict and tension, particularly with reference to those organisations engaged in the discourse of national and international environmental protest. The internment of both Aleksandr Nikitin and Grigorii Pas'ko, as a consequence of their work in relation to the nuclear safety of the Northern and Pacific navy fleets, received a great deal of media attention.[11] Nevertheless, attempts to force a national referendum during 2000 in response to the abolition of *Goskomekologiya* and the Federal Forest Service as well as the import of spent nuclear fuel, while ultimately unsuccessful, appeared to demonstrate the marked regional scope of the professionalised elements of the movement, and their ability to coordinate activities on a large scale (Peterson and Bielke, 2001, p. 69). An All-Russian Conference for "Green" Movements and Civil Society, held in Moscow during October 2003, brought together over 100 socio-ecological movements from 42 federal regions in order to discuss state policy and consider ways forward for the movement as a whole (Deklaratsiya, 2003). In particular, they noted the need for developing a more effective partnership between the 'green' movement and state organs. More generally, they outlined the importance of purposeful engagement with business activities and the strengthening of relations between environmental groups, both domestically and at the supranational scale. At the same time, such activities should not be taken as evidence of widespread environmental awareness and activism within Russian

---

[11] See the Bellona Foundation website for more information on Nikitin and Pas'ko, http://www.bellona.no/.

society (e.g. see the local level work of Crotty, 2003). The state has made some attempts to encourage greater public involvement in government decision-making; however, the extent to which such efforts have been successful is not entirely clear. Initiatives specific to fostering dialogue between the state and environmental groups are reported yearly in the State of the Environment Report, and yet there is an overall sense that progress is slow and inhibited by a range of entrenched institutional factors and vested interests (e.g. see Mazurov, 2004, p. 15). Furthermore, weak financing and a lack of basic technical infrastructure are main constraining factors for large sections of Russia's environmental movement. While a lack of empirical work makes it difficult to determine the emerging dynamics between environmental groups and corresponding state structures, it would appear that the larger organisations are engaging with federal structures and have some influence on the development of state policy. Nevertheless, this influence should not be overstated, and there is clear scope for more fruitful interaction between these two groups in both the formation and implementation of policy initiatives.

Concluding Remarks

Russia inherited a reasonably extensive system of environmental governance from the Soviet period characterised by sensibilities that had overlap with those concretised at the international level during the Rio process of the 1990s. For example, there was an emphasis on improving efficiency levels and reducing pollution output within the national economy, and this dovetailed with concerns embedded in the international discourse surrounding the concept of sustainable development (see also Chapter 2). Nevertheless, the fundamental changes taking place within Russian society more generally at this time necessitated a similar radical reworking of the country's environmental governance structures. The resulting process of reorganisation has taken many guises and included an overhaul of related legislation in order to work through more purposefully the environmentally-oriented articles outlined in the 1993 Constitution and, to a lesser extent, the central tenets of sustainable development. Furthermore, environmental legislation has also been reworked in order to accommodate and regulate the country's developing market-type relations. Nevertheless, changes made to Russia's environmental administrative framework are suggestive of a marked shift in the status and role of environmental protection functions within Russia's executive organs of power. During the mid to late 1990s, environmental concerns maintained a more or less dominant role vis-à-vis natural resource use. However, successive administrative reorganisations have witnessed the ascendancy of natural resource interest groups and a concomitant reduction in the independence and effectiveness of environmental protection and monitoring organs. For some, the emergence of a super ministry, encompassing all areas of natural resource appropriation, is a necessary development providing the basis for a more coordinated and comprehensive approach to resource utilisation. This centralising tendency also reflects wider trends evident within Russia's administrative structures. Furthermore, there is little doubt that natural resource management in certain regions of Russia would benefit from greater levels of coordination (see Chapter 5). However, the current institutional arrangement is suggestive of a conscious attempt to downplay the potentially obstructive pronouncements that might emanate from an independent environmental body, which would have implications for Russia's short- to medium-term resource-use strategy. It would also appear to signal the emergence of more overt recognition by central state organs of the primary role Russia's natural resource wealth plays in the country's economy. In response to this, some commentators have viewed the gradual erosion of independent environmental protection powers within the administration as symbolic of a purposeful 'de-ecologisation' of state policy (e.g. Danilov-Danil'yan and Yablokov, 1999; Deklaratsiya, 2003).

As suggested above, the ongoing reformulation of legislative and administrative structures cannot be understood without recourse to a range of influences. These include the concept of sustainable development, related international environmental discourse as well as more general domestic political-economic concerns. However, it is difficult to determine an objective assessment concerning the effectiveness of Russia's evolving system of environmental governance, which accounts for the myriad of processes operating at both domestic and global scales. In his work related to environmental trends in the Czech Republic post-1989, Fagin (2001) employs the concept of environmental capacity in order to explore the ability of the country to address its environmental issues. While acknowledging the concept's usefulness in highlighting policy and structural shortcomings, Fagin is critical of its ability to theorise adequately the complexities of the post-socialist societies in CEE. This is particularly relevant in recognition of its tendency to draw from Western experience and encourage a 'check-list' approach to evaluating recent changes similar to that implicit within so much economic assessment. By extrapolation, such an approach can underestimate, or indeed ignore, the environmental capacities characteristic of the late Soviet era as well as the associated implications of these capacities for the contemporary period.

Bearing this in mind, it would appear reasonable to conclude that Russia's emerging system of environmental governance has been undermined consistently by a number of factors. First, the growing importance of natural resource exploitation and use in relation to environmental protection functions has been a noticeable theme since the mid to late 1990s. Second, the evident gap between policy rhetoric and effective action is evident in many areas of environmental policy. Related to this is the limited availability of financial resources leading to the poor implementation of action programmes and a weakening of monitoring and regulatory activities. Third, while it is important to recognise the positive aspects of Soviet environmental practices, it is also necessary to acknowledge that this period was characterised by obstructive policies and institutional arrangements, which were maintained to varying degrees after 1991. A fourth factor concerns the scope of both legislative reworking and institutional reorganisation since 1991. Such an extensive process of restructuring within the context of deep societal change generates multiple contradictions in the evolving governance structures and policy formulations. This undermines overall efficiency levels and provides a window of opportunity for environmental crime and related activities. At a more general level, it can be hypothesised that the uncertainty surrounding the nature of the state administrative structure also helps to weaken the effectiveness of environmental protection activities in other ways. In particular, if environmental protection activity is understood as a process grounded in the day-to-day interaction between the state regulatory body (operating at a range of scales) and those entities requiring regulation, such as industrial units, businesses and agricultural concerns, then the constant rearrangement and redeployment of personnel can do little to enable the building of effective working relationships (see Pickvance, 2003). Similarly, the recent reconfiguration of the Ministry for Natural Resources has had repercussions for international cooperation due to the disruptive nature of the restructuring process (e.g. European Commission, 2001a, p. 10).

Chapter 5

The Changing State of the Russian Environment

Introduction

The main aim of this chapter is to provide an overview of Russia's contemporary environmental situation, highlighting trends at the national and regional level since 1991. As noted in Chapter 1, the intention is not to give an exhaustive account of different pollution types, but rather to provide a feel for changes in the underlying state of the environment since the fall of the Soviet Union. More specifically, this chapter extends the general analysis carried out in Chapter 3, with the purpose of examining general pollution trends at both the national and regional level, in addition to associated changes in environmental quality. The precipitous falls in gross pollution levels since 1991 encourage a largely apathetic approach towards Russia's contemporary environmental issues within many quarters. There is the assumption that while the overall state of the environment may not be improving, it is certainly not deteriorating. However, an appreciation of prevailing trends would suggest that such apathy is misplaced. Indeed, the more detailed analysis provided below highlights a dynamic situation which is not reducible to the simple rhythms of gross pollution output and economic restructuring. Furthermore, it is important to note at this juncture that Russia remains a key global environmental actor in spite of the scope and scale of societal restructuring in recent years and related falls in gross levels of pollution.

Before moving on to the substantive focus of the chapter, the following section is devoted to an exploration of Russia's environmental monitoring system in order to gain insight into the way in which environmental data are generated and presented in official documentation.

Environmental Monitoring Procedures

Mention has already been made of the relatively extensive monitoring network that Russia inherited from the Soviet period, a network that had developed substantially during the last two decades of the regime. For example, while air quality monitoring was carried out in just 45 urban regions during the mid-1960s, by 1989 *Gosgidromet*'s (State Hydrometeorological Service) network of monitoring stations encompassed 550 urban regions across the former Soviet Union (Kostenchuk et al., 1993, section 2.1.1; see also Bridges and Bridges, 1995). In addition, Bridges (1992, p. 257) draws attention to the extent of the Soviet Union's water monitoring system during the late 1980s, consisting of more than 3500 surface water-monitoring sites. The Soviet period also left a legacy of comparatively stringent environmental quality standards. During the mid-1990s, Russian air quality thresholds for a range of pollutants were typically more demanding than those utilised by the EU and laid down by the World Health Organisation (WHO), although these were not necessarily adhered to in practice (see OECD, 1999, pp. 63-64). At a general level, this strictness can be related to the ideological underpinnings of the Soviet system, which prioritised the ability of the centralised state structure to maintain a favourable environment in order to ensure the health of the Soviet population. Nevertheless, in spite of these positive characteristics, the effectiveness of the Soviet Union's monitoring system was undermined by a lack of integration between its constituent parts, typically reducible to bureaucratic and administrative obstacles (IMF et al., 1991, pp. 15-16; OECD, 1996; see also Chapter 4). For example, the branch logic of the prevailing administrative system, combined with the weakness of centralised environmental control functions relative to industrial ministries, ensured that information was not necessarily utilised in the most effective manner (see Chapter 2). Many of these noted weaknesses continued to characterise Russia's monitoring system as it emerged post-1991. In order to address the evident need for greater coordination of monitoring activities, and also to ensure comparability with international norms, the Russian government initiated the development of a Unified State System for Environmental Monitoring (USSEM) during the early 1990s (see MEPNR, 1994, pp. 208-213).[1] However, work related to the monitoring system has taken place unevenly across Russian territory (e.g. see OECD, 1996) and, at the same time, overall progress has been substantially undermined by continuous administrative reorganisation and financial shortfalls (MPR, 2002, p. 341).

These problems have had a detrimental effect on the efficiency of monitoring networks associated with key organisations such as the Federal Service for Hydrometeorology and Monitoring of the Environment (*Rosgidromet*) and the Ministry of Health's Federal Centre for Sanitary Epidemiological Inspection (*Gossanepidnadzor*). *Rosgidromet* (based on the organisational structure of the Soviet Union's State Hydrometeorological Service) operates an extensive monitoring network and produces a wide range of environmental data. At the same time, the organisation was not immune to the disruptions of the early post-Soviet period, and limited funding during the mid-1990s placed its monitoring network under considerable pressure (e.g. Smol'yakova, 1997, p. 4). Elements of *Rosgidromet*'s monitoring network appeared to contract during the 1990s and, while there has been some recovery in the network's capacity during recent years, it is likely that the comparability of data-sets across the years has been undermined (see Table 5.1). This situation is replicated at the local level. For example, the number of stationary posts comprising Moscow city's air pollution monitoring network fell from a mid 1980s high of 60 to less than 20 by the mid 1990s (see Oldfield, 1999). The sanitary-epidemiological monitoring activities of *Gossanepidnadzor* (which according to Bridges and Bridges (1995, p. 141) date

---

[1] The formation of the USSEM was initiated by government decree No. 1229, 24/11/1993.

back to the early post-war years) undoubtedly suffered similar disruption during the course of the 1990s.

Table 5.1 Changes in the extent of *Rosgidromet*'s pollution monitoring system according to selected indicators, 1992-2003

Type of monitoring	Extent of monitoring activity			
	01/01/1992	01/01/1994	01/01/2002	01/01/2004
Regular air quality testing (urban regions)	255 cities and settlements	248 cities and settlements (682 stations)	225 cities and settlements (618 stations)	229 cities and settlements (623 stations)
Soil pollution (agricultural regions)	300 farms -	268 farms (160 regions)	- (190 regions)	- (190 regions)
Inland surface water pollution	Hydro-chemical monitoring: 1194 streams & 147 reservoirs (natural and artificial) Hydro-biological monitoring: 196 water bodies	Hydro-chemical monitoring: 1175 streams & 151 reservoirs (natural and artificial) Hydro-biological monitoring: 190 water bodies	Hydro-chemical monitoring: 1195 water bodies (includes streams & reservoirs) Hydro-biological monitoring: 107 water bodies	Hydro-chemical monitoring: 1182 water bodies (includes streams & reservoirs) Hydro-biological monitoring: 133 water bodies
Snow cover pollution	Approx. 640 points	625 points	565 points	536 points
Sea pollution	Hydro-chemical/hydro-biological monitoring: littoral areas of 11 seas	Hydro-chemical/hydro-biological monitoring: littoral areas of 11 seas	Hydro-chemical monitoring: littoral areas of 8 seas	Hydro-chemical monitoring: littoral areas of 8 seas

- Data not available

Sources: compiled from MinEkologiya, 1992, pp. 65-66; MEPNR, 1994, pp. 211-212; MPR, 2002, p. 345; MPR, 2004, section 7.

Statistical Reporting of Environmental Data[2]

The State Statistical Committee (*Goskomstat*) is responsible for publishing a range of environmental volume data both in its yearly reports and in less regular publications devoted to environmental protection. These data are generated via the returns from self-reporting enterprises and organisations (Dumnov, 2002, p. 55). In addition to the statistical forms associated with the activities of *Goskomstat*, a plethora of other statistical data is derived via the accounting activities of individual ministries and other administrative organs. In total there are approximately 200 statistical surveys in existence relating to a wide range of natural resource and environmental protection characteristics. In the case of *Goskomstat*, approximately 40 statistical surveys are currently carried out (typically on a yearly basis) and these can be subdivided into 12 themed sub-sections (Table 5.2). It should be noted that the Soviet self-reporting system was not exhaustive of all polluting sources with various industrial units absent from the underlying database (e.g. see Peterson, 1993, p. 21). Such deficiencies undoubtedly continue to be a feature of the post-Soviet situation. An analysis of reporting procedures during the 1990s indicates that pollution data also ignore a number of significant sources of air and water pollution such as emissions from domestic heating appliances (CIS STAT, 1996, p. 186), irrigation discharge water and urban run-off (Bridges and Bridges, 1996, pp. 73-74).[3] The comprehensiveness of *Goskomstat's* reporting system has been questioned further in recent years due to the marked structural changes taking place within Russian society. For example, it would seem likely that many new economic entities and forms of economic activity remain outside of any formal assessment procedure and are thus not incorporated effectively within the overall accounting system. At the same time, much of this new activity operates at a small-scale and is oriented to the developing service sector and therefore responsible for relatively small volumes of pollution emissions (Dumnov, 2002, p. 51).

While financial difficulties have compromised monitoring networks at both the national and local level, competing external pressures have resulted in the internalisation of certain international environmental accounting procedures. The European Union has attempted to encourage Russia's conformity with its own environmental standards via the dissemination of strategic funds through TACIS assistance (European Commission, 2003, Annex 1, p. vi). New areas of environmental reporting have also been opened up with the assistance of foreign sponsors. For example, work carried out by organisations such as the OECD during the course of the last decade has helped to establish an approximate understanding of Russia's municipal waste problem (see Chapter 3). Furthermore, inventories related to greenhouse gas emissions data, as well as reviews of biodiversity resources, have evolved during the course of the 1990s in liaison with the corresponding secretariats of the UNFCCC and Biodiversity Convention (see Amirkhanov, 1997; Hanna et al., 2000, p. 27). Importantly, Russia's ability to meet the requirements of such conventions has often been undermined by prevailing socio-economic difficulties. As a consequence,

---

[2] See also Shaw and Oldfield (1998, pp. 165-167).

[3] The author would like to thank Nathaniel Trumbull for additional insight concerning the inadequacies of contemporary pollution data.

organisations such as the Global Environment Facility (GEF) and the United Nations Development Programme (UNDP) have provided necessary funds to facilitate an effective response.[4]

Finally, it is worth making a few points about the presentation of environmental data in government publications. It is the convention in much of the official literature to publish environmental quality data as a function of the maximum permissible concentration (*predel'no dopustimaya kontsentratsiya*) or other norms. The absence of absolute indicators of pollution levels within published volumes makes it difficult to undertake international comparisons and can also mask changes in environmental quality where concentration levels change but remain below the established limits. Russia's noted tendency to set relatively strict sanitary limits in some cases should also be acknowledged when dealing with data-sets based on exceeding the threshold limit. This publishing convention, in conjunction with the inconsistencies extant within Russia's monitoring network and self-reporting system, ensures that data should be treated with caution. At the same time, a judicious engagement with data trends over the course of the last decade is capable of revealing some of the main underlying features and trends of Russia's contemporary environmental situation.

Table 5.2 Categories utilised by *Goskomstat* for generating statistical data related to nature-use and environmental protection

Main sub-categories

Geological survey work and extraction of economic minerals
Forestry activity and specially protected natural territories
Utilisation and protection of water resources
Protection of air resources
Biological resources
Formation and utilisation of waste and secondary energy resources
Urban municipal infrastructure
Accidents and emergency ecological situations
Investment in environmental protection
Scientific research and innovation
Higher education and professions
Criminal activity related to natural resources and nature protection

Source: compiled from Dumnov, 2002, pp. 56-59.

---

[4] For example, GEF provided a grant of US$ 53 thousand to enable Russia to produce a first national report on Biodiversity as part of its commitment to the Convention on Biodiversity (www.gefonline.org).

Air Pollution Trends Post-1991

Introduction

While the Soviet Union generated significant levels of air pollution, gross pollution levels were not excessive in comparison with other large industrial countries such as the USA. For example, Chapter 2 noted that the Soviet Union was producing approximately 75 percent of the USA's total air pollution output at the end of the 1980s, although data inadequacies suggest that this probably underestimated the USSR's total output. At the same time, it was indicated that levels of pollution intensity were greater in the Soviet case, with pollution per unit of GDP far in excess of that reported for the USA. Russia's contemporary pollution profile continues to reflect these general trends, being characterised by significant overall levels of emissions and comparatively high levels of pollution intensity. This is illustrated in Table 5.2, which compares Russia's air pollution emissions for key pollutants with a range of OECD countries. The USA is a clear leader in terms of overall volume discharges for all four of the listed traditional air pollutants. Russia and Australia are also highlighted as significant polluters but both lag someway behind the output levels of the USA. Once pollution emissions are calculated on a per capita basis, Russia loses its dominant position whereas other large gross polluters such as the USA, Australia and Canada retain their pre-eminence. The final column of Table 5.2 relates pollution output to levels of GDP and, as expected, Russia performs poorly in this respect exhibiting relatively high returns for all four pollutants. Nevertheless, countries such as Australia and Canada display figures of a similar magnitude. The variation in systems of data generation and coverage ensures that comparative analysis between countries is an imprecise science. Suffice to say that Russia's air pollution emissions, while certainly significant at the global scale, are comparable in extent to those being produced by other large industrial countries. Table 5.3 provides a useful backdrop to the following discussion related to the specifics of Russia's air pollution output since 1991. In particular, it helps to refocus attention on Russia's continuing importance at the global level as a producer of air pollution emissions despite the recent marked declines in stationary source air pollution emissions.

Stationary Source Air Pollution

Chapter 3 highlighted the extent to which the level of pollution emissions fell during the 1990s as economic production collapsed (see Figure 3.1). In the case of stationary source air pollution emissions, the percentage decline was indeed significant, with output levels for 2001 just 60 per cent of those in 1990, and this on top of the noted declining trend during the 1980s. As such, Russia's SSAP output during the early part of the 21st Century is significantly lower than was the case at the end of the Soviet period. At the same time, levels of SSAP increased by 1.3 million tonnes (7 percent) between 1999 and 2003 reflecting Russia's strengthening economic performance and it is likely that this impetus will be maintained over the short- to medium-term (Goskomstat, 2004b, p. 20). As noted

Table 5.3 Comparative emission data of traditional air pollutants (both stationary and mobile sources) for Russia and selected OECD countries, late 1990s

	Total emissions (000s tonnes)				Per capita emissions (kg/cap)				Emissions per unit of GDP (kg/000 US$)			
	SOx^b	NOx^c	CO^d	VOC^e	SOx	NOx	CO	VOC	SOx	NOx	CO	VOC
Australia	1818	2565	18166	1917	95.8	135.2	957.8	101.1	4	5.7	40.1	4.2
Canada[f]	2691	2056	10145	2670	89.7	67.4	338.3	89	3.7	2.6	14	3.7
Czech Republic[f]	265	397	649	247	25.8	38.6	63.2	24	2	3	4.9	1.9
Germany	831	1637	4952	1651	10.1	19.9	60.3	20.1	0.4	0.9	2.7	0.9
Hungary[f]	592	221	737	149	58.5	22	72.9	14.8	5.7	2.1	7.2	1.4
Japan	870	1654	3636	1842	6.9	13.1	28.7	14.5	0.3	0.5	1.2	0.6
Poland[f]	1511	838	3463	599	39.1	21.7	89.6	15.5	4.3	2.4	9.9	1.7
Russian Federation	*5877*	*3029*	*16225*	*3019*	*39.9*	*20.5*	*110*	*20.5*	*6*	*3.1*	*16.7*	*3.1*
UK	1187	1603	4760	1566	19.9	26.9	80	26.3	1	1.3	3.9	1.3
USA[f]	17116	23037	85648	15763	62.7	84.4	313.8	57.8	2	2.6	9.8	1.8
OECD average[a]	36700	44900	170400	37000	32.9	40.3	152.8	35.1	1.5	1.9	7.1	1.6

Note that the original OECD source data source advises caution when making cross-country comparisons due to differences in data generation.

[a] OECD data are compiled from the following countries: Canada, Mexico, USA, Australia, Japan, South Korea, New Zealand, Austria, Belgium, Czech Republic, Denmark, Finland, France, Germany, Greece, Hungary, Iceland, Ireland, Italy, Luxemburg, Netherlands, Norway, Poland, Portugal, Slovak Republic, Spain, Sweden, Switzerland, Turkey, UK.
[b] SOx – Sulphur oxides
[c] NOx – Nitrogen oxides
[d] CO – Carbon monoxide
[e] VOCs – Volatile Organic Compounds
[f] SOx figure refers to SO_2 emissions only

Source: compiled from OECD, 2002a, pp. 6-7.

previously, the particularities of the Soviet Union's development model ensured that industrial activity was concentrated in specific urban regions, thus giving rise to a marked asymmetrical distribution of pollution load within the urban hierarchy and this is most obvious with regard to SSAP. In 1990, over 45 percent of Russia's total discharge of SSAP was attributable to just 10 constitutive regions.[5] These

---

[5] During the late Soviet period, the Russian Soviet Federative Socialist Republic (RSFSR) consisted of 70 constituent parts (including autonomous Soviet Socialist Republics, autonomous *oblasti* and autonomous *okruga*). The Russian Federation is comprised of 89 federal units and this total includes the city regions of Moscow and St. Petersburg as well as Kaliningrad *oblast'*. There is sufficient continuity between these two periods to enable statistical comparisons to be made.

tended to be located in the heavy industrial belts of the middle Volga, the southern Urals and the southern regions of Siberia (see Figure 5.1). Within the context of significant falls in the overall level of SSAP during the 1990s, this pattern of dominance appeared to intensify, to the extent that in 2002 the 10 most polluting federal regions were responsible for approximately 60 percent of the Russian total. Furthermore, there was little change in the geography of major polluting regions during this period, reflecting the inertia of previous patterns of industrial production (Figure 5.2). Nevertheless, within this general trend it is worth noting the considerable increase in SSAP output attributable to the Khanty-Mansi autonomous *okrug* when comparing 2002 with 1993 data (+88 percent), testimony to this region's importance for Russia's hydrocarbon production output. The adjacent autonomous region of Yamalo-Nenets is also a key hydrocarbon producing region, and a similar upward trend is noted when comparing data for the two years (+48 percent). The only other region amongst those highlighted in Figures 5.1 and 5.2 recording a higher level of SSAP output in 2002 than at the beginning of the 1990s was Kemerovo *oblast'* located in south-east Siberia (+1.5 percent, 2003 as a percent of 1990).

Mobile Source Air Pollution

Air pollution emissions attributable to motor vehicles (mobile source air pollution [MSAP]) have followed a different trajectory to that of SSAP. According to official data-sets, emission levels fell markedly between 1990 and 1995 (approximately 50 percent), before recovering slowly in recent years. Year on year increases since 1995 have resulted in a 30 percent rise in mobile source air emissions through to 2002, although the figure of 14.4 million tonnes is little more than two-thirds of emission levels in 1990 (MPR, 2003, p. 189; Goskomstat, 2001a, p. 21). Runs of data collated by the OECD for individual pollutants (1990-1998) provide additional insight into this trend (OECD, 2002a, pp. 21, 33). For example, the emission levels for nitrogen oxides (NOx) attributable to mobile sources of pollution increased by more than 20 percent over the period 1990-1998. In the case of carbon monoxide (CO) emissions, output attributable to motor vehicles fell in the early part of the 1990s, but then increased year on year between 1995 and 1998. While there is uncertainty concerning the accuracy of the available data, particularly in recognition of the recent marked increase in private car use, the initial contraction in mobile source emissions during the early 1990s can be related to falling levels of economic activity and production output. In contrast, recent increases are correlated strongly with the substantial expansion of private vehicle numbers throughout the country's main urban regions. Numbers of private passenger vehicles more than doubled during the period 1990 to 2001, rising from approximately 9 million to over 22 million (Goskomstat, 1996, p. 594; MPR, 2003, p. 190).

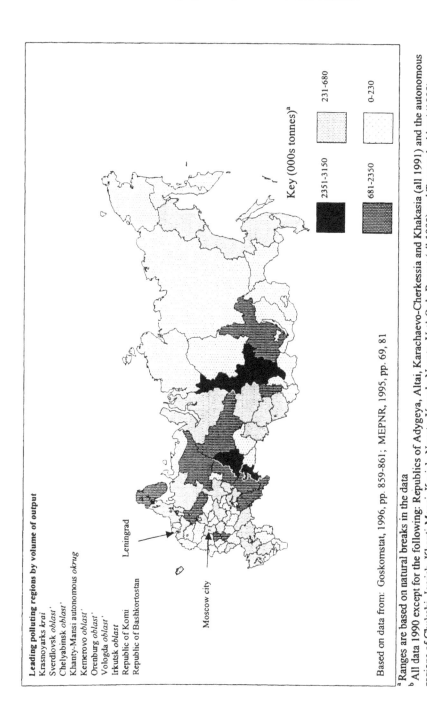

Leading polluting regions by volume of output

Krasnoyarks *krai*
Sverdlovsk *oblast'*
Chelyabinsk *oblast'*
Khanty-Mansi autonomous *okrug*
Kemerovo *oblast'*
Orenburg *oblast'*
Vologda *oblast'*
Irkutsk *oblast*
Republic of Komi
Republic of Bashkortostan

Leningrad

Moscow city

Key (000s tonnes)[a]

2351-3150
681-2350
231-680
0-230

Based on data from: Goskomstat, 1996, pp. 859-861; MEPNR, 1995, pp. 69, 81

[a] Ranges are based on natural breaks in the data
[b] All data 1990 except for the following: Republics of Adygeya, Altai, Karachaevo-Cherkessia and Khakasia (all 1991) and the autonomous regions of Chukchi, Jewish, Khanti-Mansi, Koriak, Nenets, Yamalo-Nenets, Ust'-Orda Buryat (all 1993) and Tyumen' *oblast'* (1993).

Figure 5.1   Stationary source air pollution emissions according to federal region, 1990

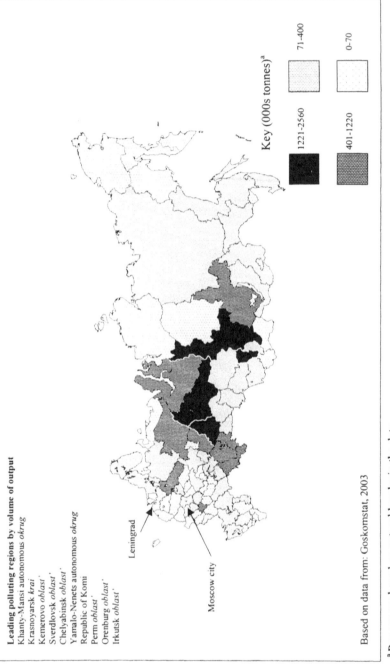

Leading polluting regions by volume of output
Khanty-Mansi autonomous *okrug*
Krasnoyarsk *krai*
Kemerovo *oblast'*
Sverdlovsk *oblast'*
Chelyabinsk *oblast'*
Yamalo-Nenets autonomous *okrug*
Republic of Komi
Perm *oblast'*
Orenburg *oblast'*
Irkutsk *oblast'*

Leningrad

Moscow city

Key (000s tonnes)[a]

1221-2560
401-1220
71-400
0-70

Based on data from: Goskomstat, 2003

[a] Ranges are based on natural breaks in the data

Figure 5.2   Stationary source air pollution emissions according to federal region, 2002

As a consequence of the emission trends for both SSAP and MSAP, the mix of air pollution discharges according to source has changed during the last decade at the national level. This change is relatively small but nevertheless suggestive of a concomitant change in the mix of air pollutants present in urban regions. The following section provides support for this trend. At present, motor vehicle pollution dominates air pollution output levels in a number of large urban areas. Moscow is the most obvious example with motor vehicle emissions responsible for approximately 90 per cent of total air pollution output (see Bityukova and Argenbright, 2002; MPR, 2002, p. 98; Oldfield, 1999).

(Photograph: author)

Plate 5.1   Garage developments (to accommodate Moscow's burgeoning private car fleet) encroaching on green spaces amid apartment blocks

Air Quality

The marked change in the level of volume pollution emissions during the 1990s, in tandem with the structural changes within Russian society, has implications for the quality of air resources. According to *Rosgidromet*, the concentration of urban-based pollutants such as suspended substances, sulphur dioxide, ammonia, phenol, soot and carbon bisulphide fell at the national level during the period 1990-1999 (Table 5.4). Such trends are related predominantly to the contraction in levels of industrial production and related activity. Air quality data are also generated by *Gossanepidnadzor* as part of its more general remit to monitor the influence of environmental factors on the health and well-being of the population. Available data are indicative of an improving situation at the national level during the period 1991 to 2001. For example, over 10 percent of sanitary tests carried out in 1993 exceeded the maximum permissible concentration (MPC) (Gossanepidnadzor, 1995, p. 13) compared to approximately 6 percent in 2001 (MPR, 2002, p. 98).

Table 5.4 Percentage change in the average concentration of main air pollutants in Russia's urban areas, 1990-1999 and 1999-2002

Pollutant	% change in average yearly concentration, 1990-1999	% change in average yearly concentration, 1999-2002
Benz(a)pyrene[a]	-55	+47
Sulphur dioxide[b]	-49	-4
Carbon bisulphide[c]	-45	-
Ammonia[d]	-34	-
Suspended substances[e]	-18	-6
Soot[f]	-17	-
Phenol[g]	-7	-
Hydrogen flouride[h]	-5	-
Nitrogen oxide[i]	-5	+10
Formaldehyde[j]	+1	-
Hydrogen sulphide[k]	+5	-
Nitrogen dioxide[i]	+13	-4
Carbon monoxide[l]	+15	-3

- Data not available
[a] Sources include non-ferrous metallurgy, fuel combustion and road transport.
[b] Sources include power generation and related industrial activities.
[c] Sources include industrial processes, landfills and natural gas production/distribution.
[d] Sources include intensive agricultural livestock practices.
[e] Sources include both industrial and transport sources.
[f] Sources include industrial sources. Associated with black smoke.
[g] Sources include heavy industrial activities (e.g. oil refinery, metallurgy) and municipal waste treatment.
[h] Sources include fuel combustion activity.
[i] Sources include transport and industrial combustion processes.
[j] Sources include industrial processes, fuel combustion and transport.
[k] Sources include fuel combustion, hydrocarbon processing, sewage treatment, transport.
[l] Sources include transport and fuel combustion processes.

Note: For detailed information concerning individuals pollutants and their possible sources see: UK's National Atmospheric Emissions Inventory: http://www.naei.org.uk/index.php, Australia's national database of pollutant emissions: http://www.npi.gov.au/index.html, and the US Environmental Protection Agency: http://www.epa.gov/.

Sources: Goskomekologiya, 2000, Table 1.1; MPR, 2003, p. 9.

While such data are supportive of an improving trend in air quality at the national level for a range of main pollutants, it is important to recognise that the situation can be far more complex at the regional and local level. Indeed, air quality remains problematic for millions of urban-based Russians, with the most polluted urban regions typically associated with heavy industrial activity such as ferrous and non-ferrous metallurgy and the oil and petro-chemical industry (MPR, 2002, p. 11). Furthermore, the national trend for certain key pollutants since the mid-late 1990s is characterised by a worsening situation. This is linked, in certain instances, to a combination of Russia's strengthening economy and general infrastructural weaknesses, giving rise to inefficient industrial processes and accidental releases of polluting substances. However, the positive data trends outlined in Table 5.4 are primarily reflective of the growing size of Russia's motor vehicle fleet, with nitrogen oxides, carbon monoxide, hydrogen sulphide and formaldehyde all having links, to a greater or lesser extent, with exhaust emissions. For example, average urban concentrations of key exhaust emissions, such as carbon monoxide and nitrogen dioxide, increased markedly at the national level during the 1990s. Furthermore, nitrogen oxide recorded an increase of +10.3 percent over the period 1999-2002.[6] Benz(a)pyrene also registered a sharp increase in average yearly concentration levels between 1999-2002 after a pronounced fall during the 1990s. This pollutant has carcinogenic properties and is related strongly to metallurgical industrial processes. The relatively robust performance of this industry during recent years would seem to account for the recent upward trend in concentration levels (see also OCED, 1999, p. 58).

Additional measures of air quality utilised in the yearly State of the Environment reports include the reporting of extreme air pollution events (i.e. when a given pollutant registers more than 10 times the acceptable sanitary limit in a particular urban region) and the atmospheric pollution index (*indeks zagryazneniya atmosfery* or IZA). The IZA is comprised of air quality data (weighted for toxicity) for a range of main pollutants specific to each urban region and, as a device for assessing air quality, has its roots in the Soviet period (see Peterson, 1993, p. 38). Urban regions recording an index score of >7 are considered problematic and those with scores of ≥ 14 are deemed to be suffering 'very high' levels of air pollution. Notable increases in the number of urban localities recording scores of 7 or greater since the late 1990s suggest that air quality is becoming increasingly strained within a range of Russia's urban centres. For example, in 2002, 130 urban centres (accounting for more than 40 percent of Russia's population) registered an index score of >7 (MPR, 2003, p. 9). In contrast, 81 urban centres were included in this category in 1999. Similarly, while the number of urban centres recording a score of ≥14 fell from a high of 45 in the mid-1990s to 22 in 1999, by 2002 the figure had climbed back to 35 (Minpriroda, 1996, p. 9; MPR, 2003, p. 10). Once again there are suggestions of inconsistencies in data collection, which ensure that yearly comparisons should be treated with caution.

---

[6] Nitrogen oxide (NO) is a primary pollutant emitted directly from exhaust emissions while NO_2 is formed as NO oxidises.

Regional Trends in Russia's Air Quality

An analysis of the different air quality measures mentioned above reveals a distinctive regional geography. In particular, it implies that the Central, Siberian, and Volga federal *okruga* are characterised by a relatively high number of urban regions registering marked levels of air pollution (Table 5.5). This corresponds broadly with the regional distribution of SSAP outlined in Figures 5.1 and 5.2, as well as the generalised regional framework detailed in Chapter 3. More specifically, Irkutsk *oblast'* (Siberian federal *okrug*) is revealed as the federal unit with the highest concentration of urban regions displaying comparatively high levels of air pollution. This *oblast'* incorporates major urban localities such as Irkutsk (administrative centre) and Bratsk, as well as the smaller industrialised urban regions of Zima and Shelekhov. The region's relatively high levels of pollution are related to industrial activities (e.g. fuel-energy, non-ferrous and petro-chemical sectors), the cumulative effect of small-scale (domestic) boiler usage and the aggravating influence of prevailing climatic conditions (e.g. MPR, 2003, p. 303). Furthermore, a noted increase in the concentration of certain pollutants during recent years is blamed on a combination of increased industrial production and the accidental release of polluting substances due to technical weaknesses.

Figure 5.3 indicates those urban centres recording atmospheric pollution index scores of 14 or more in 2002. The urban regions included within the list are characterised by considerable diversity, incorporating the likes of Moscow city at the top of the urban hierarchy (approximately 9 million population) and Selenginsk (east Siberia) at the bottom (approximately 16 thousand population). A number of distinct clusters exist which tend to reflect Russia's underlying industrial geography. For example, within the Siberian federal *okrug*, concentrations of urban regions characterised by relatively high levels of air pollution are apparent in the south-eastern region (incorporating Irkutsk and the adjacent regional administrative centres of Ulan-Ude and Chita) and in the southern parts of central Siberia (encompassing the urban regions of Tomsk, Kemerovo, Novokuznetsk, Biisk and Barnaul). Clusters are also noticeable in the southern Urals region (Kurgan, Ekaterinburg, Krasnotur'insk and Magnitogorsk) and central parts of European Russia. Two additional clusters are found in the North Caucasus region and, to a lesser extent, southern parts of the Russian Far East. Noril'sk represents an obvious outlier, although Magadan and Petropavlovsk-Kamchatskii are similarly isolated. The fact that only Nizhny Novgorod from the Volga industrial region is represented in Figure 5.3 should not disguise the fact that this region contains a number of urban areas characterised by significant levels of air pollution (see Table 5.5).

Table 5.5 Air pollution parameters for urban regions, 2002

Federal *okruga*	Average yearly concentration of polluting substances exceeds MPC[a]		Atmospheric pollution index[b] >7		Atmospheric pollution index >14	
	No. of urban regions	% of total	No. of urban regions	% of total	No. of urban regions	% of total
Central	32	16	23	18	3	9
Moscow city & Moscow oblast'	*10*	*5*	*7*	*5*	*1*	*3*
North Western	23	11	10	8	1	3
Southern	19	10	12	9	7	20
Volga	42	21	33	25	1	3
Ural	17	9	11	8	4	11
Siberian	47	23	28	22	13	37
Irkutsk oblast'	*14*	*7*	*7*	*5*	*4*	*11*
Far Eastern	21	10	13	10	6	17
Total	201	100	130	100	35	100

[a] MPC – Maximum permissible concentration. It is presumed that the number and range of pollutants included within this figure varies from urban region to urban region.
[b] Air pollution index – see text for explanation

Source: compiled from MPR, 2003, pp. 12-14.

Bearing in mind the data issues outlined above, it is perhaps prudent to conclude that while Russia's air quality has improved with respect to certain pollutants, the combination of increasing motor vehicle emissions, the recent upturn in industrial production and persistent infrastructural weaknesses (e.g. see MPR, 2002, p. 146) ensures that air quality remains strained in a large number of urban regions and with a tendency to worsen in recent years. As suggested above, the overall decline in air pollution emissions during the 1990s was so extensive that worsening emission and quality trends are likely to attract limited attention both domestically and internationally, at least in the short-term. Furthermore, limited political will is evident to support the prioritisation of environmental concerns with economic recovery still in its early stages.

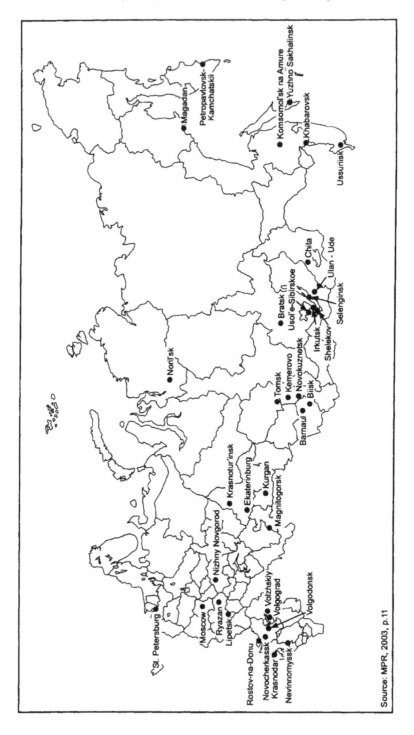

Source: MPR, 2003, p.11

Figure 5.3   Urban centres recording Atmospheric Pollution Index scores ≥ 14, 2002

Russia's Water Resources

Water Demand and Drainage Discharge

Towards the end of the Soviet period water quality gave cause for concern being characterised by relatively high levels of chemical and biological pollution (e.g. Kostenchuk et al., 1993, section 2.3). The fall in production levels during the 1990s was, understandably, expected to encourage an improvement in overall quality levels due to reduced pressure on the resource as well as associated storage, delivery and cleaning infrastructures. Nevertheless, the resulting situation is rather more complex.

Before moving on to examine water quality trends, it is first worth noting changes in the level of water consumption and drainage discharge at the national level. Water demand for both industrial and agricultural needs fell markedly during the 1990s, whereas demand attributed to the municipal sector remained relatively stable (IOPRR, 2001, p. 73). Nevertheless, the overall reduction in water demand (a fall of approximately 27 percent between 1990 and 2000) encouraged similar falls in the level of drainage discharge to surface waters. As indicated in Chapter 3, the national level of polluted drainage discharge fell more or less consistently over the period 1990-2003, and recorded an overall decline of approximately 32 percent (see Figure 3.3). Mirroring the general trend with regard to water consumption, polluted drainage discharge levels attributable to the industrial and agricultural sectors fell markedly during the 1990s, while the municipal sector registered only a minimal decline. As a consequence, municipal sector discharges now account for almost two-thirds of Russia's total polluted drainage discharge compared to the industrial sector's one-third. These discharge figures should be treated with caution. For example, changes to the accounting procedures (such as the re-categorisation of certain discharges of 'polluted' water) within a number of federal regions during 2001 helped to maintain the downward trend in polluted drainage discharge (e.g. MPR, 2002, p. 16).

As with SSAP, polluted drainage discharge has a marked regional distribution. In 1990, 10 of the RSFSR's constitutive regions were responsible for approximately 52 percent of the national total (see Figure 5.4). At this time, most of the significant polluting regions were located west of the Urals, along the mid-lower parts of the Volga and the North Caucasus region. The urban agglomerations of Leningrad and Moscow were also important sources of polluted drainage discharge. The major exceptions to this general trend included Krasnoyarsk *krai* and Irkutsk *oblast'*, both located in east Siberia. While this general pattern has endured in recent years (see Figure 5.5), it should be noted that the industrial regions of Chelyabinsk, Sverdlovsk (both Ural federal *okrug*) and Kemerovo (Siberian federal *okrug*) have relatively more importance than before. In addition, the federal city regions of Moscow and St. Petersburg have established themselves as the two largest polluters, with Moscow city clearly dominant in terms of the overall volume of discharge (approximately 13 percent of the national total).

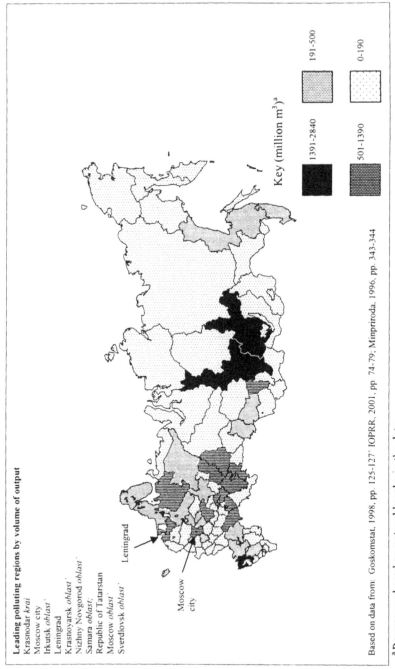

Figure 5.4 Polluted drainage discharge, 1990 (million m³)

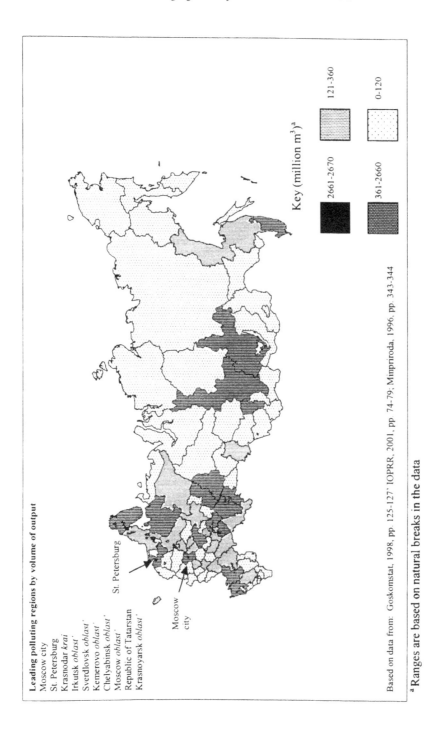

Leading polluting regions by volume of output

Moscow city
St. Petersburg
Krasnodar *krai*
Irkutsk *oblast'*
Sverdlovsk *oblast'*
Kemerovo *oblast'*
Chelyabinsk *oblast'*
Moscow *oblast'*
Republic of Tatarstan
Krasnoyarsk *oblast'*

St. Petersburg

Moscow
city

Key (million m^3)[a]

2661-2670

361-2660

121-360

0-120

Based on data from: Goskomstat, 1998, pp. 125-127' IOPRR, 2001, pp. 74-79; Mmpriroda, 1996, pp. 343-344

[a] Ranges are based on natural breaks in the data

Figure 5.5   Polluted drainage discharge, 2002 (million m^3)

From the previous discussion it follows that the efficiency of municipal infrastructure is an important element determining both the quality of urban drinking water and water sources more generally. The difficult financial situation evident at all levels of government has helped to undermine further the already strained system of water provision and cleaning inherited from the Soviet period. According to official figures, more than one third of Russia's existing water and sewage works is now in need of a radical overhaul and modernisation (e.g. MPR, 2003, pp. 183-184). Indeed, the state of disrepair is such that resources for planned overhauls are being redirected, in many instances, to address ongoing emergency situations such as pipe leakage and equipment failure (European Commission, 2003, pp. 29-30). Furthermore, these infrastructural inadequacies have implications for water quality far beyond the confines of the urban region itself.

Water Quality

Despite the fall in the volume of pollution discharge during the 1990s, official reports suggest that water quality failed to improve to any great extent (e.g. Gossanepidnadzor, 2003, pp. 14, 19; IOPRR, 2001, pp. 85). Many of Russia's water bodies continue to register pollution levels in excess of accepted sanitary norms and this is attributed in part to unregulated (*neorganizovannyi*) discharges and the infringement of sanitary protocols, particularly in rural areas (IOPRR, 2001, pp. 85-86). Furthermore, the parlous financial state of many industrial enterprises and municipal authorities undermines their collective ability to enforce strict water protection measures. As a consequence, Russia's river resources are characterised by significant pollution levels along certain sections of their course (e.g. Gossanepidnadzor, 2003, pp. 14-16; IOPRR, 2001, p. 86).

Russia is endowed with a number of major river systems stretching from the Volga system in the west to the Angara and Amur systems in the east (see Figure 2.1). Qualitative assessments carried out during the late 1990s suggest that the Volga (European Russia) and Yenisey (Siberia) river networks are characterised by particularly extensive pollution issues related to the heavy concentration of industrial activity and large urban agglomerations within their respective catchment areas. A significant number of other rivers located throughout the Russian Federation are highlighted as having marked pollution issues at certain points along their course. These include the Ural (southern Urals), Don (southern European Russia) and Neva river systems (northwest European Russia) as well as rivers located on the island of Sakhalin, the Kola Peninsula and within the Moscow region (see also Figure 2.1). There is a clear continuity here with those water issues identified towards the end of the Soviet period (Chapter 2). Furthermore, a comparison of evaluative data, covering the period 1992 to 1998, is suggestive of little or no improvement in the ecological state of a number of Russia's large rivers, and this in spite of the noted fall in pollution discharge levels (IOPRR, 2001, p. 84). A more detailed analysis of regional river pollution issues highlights the variation in underlying causal factors. For example, oil and gas extraction activity is blamed for the relatively poor sanitary state of water resources in Khanty-Mansi autonomous *okrug* whereas it is the operation of paper and pulp enterprises in Arkhangel'sk *oblast'* which is strongly associated with the pollution of the Northern Dvina river system (Gossanepidnadzor, 2003, p. 15). Where river systems incorporate large tracts of Russian territory, localised pollution issues are likely to emerge along the river course reflecting changes in economic activity, and this is evident in the case of the river Volga. According to *Gossanepidnadzor*, additional causal factors that vary regionally, and yet contribute to overall levels of water quality, include the natural state of the water source (e.g. hardness, mineral content of water etc.) and the technical state of water distribution and cleaning infrastructure (Gossanepidnadzor, 2003, p. 25).

Evaluative work carried out by *Gossanepidnadzor* during the 1990s provides further evidence of the strained situation with regard to water quality. As indicated above, *Gossanepidnadzor*'s primary role is to gather data in order to assess the influence of environmental factors on human health. As such, it carries out sanitary testing of drinking water sources and drinking water systems. While such data are useful in indicating the relative effectiveness of water cleaning infrastructure, they also provide an insight into more general water quality trends. Table 5.6 outlines sanitary-chemical and microbiological trends for sources of centralised water supply, and the water distribution system, at the national level. In general, the data are suggestive of a persistently strained sanitary state with respect to sources of centralised water supply during the course of the 1990s, alongside some improvement in the sanitary state of the water supply system itself.

Table 5.6 Sanitary testing carried out on drinking water sources and the water supply system, % of tests failing the accepted norm (national average)

	Type of sanitary test	1991	1993	1995	1996	1997	1998	1999	2000	2001	2002
Sources of centralised water supply[a]	Chemical tests		20.4		29	29	29	28.7	28.4	28.3	27.9
	Micro-biological tests		11.2		9.2	9.7	9.4	9	9.1	8.6	8.0
Water supply system[b]	Chemical tests	23.8	21.4	20.9	19.6	20.1	20.6	19.7	20.3	19.5	18.9
	Micro-biological tests	12.8	11.2	10.9	10	10.3	10.6	9.8	9.4	9.1	8.1

˙ Data not available

[a] Data refer to both surface and groundwater sources of centralised water supply.

[b] *Vodoprovodnaya set'*.

Sources: Goskomstat, 2001a, p. 24; Gossanepidnadzor, 1995, pp. 11-15; 2001, pp. 17; 2003, pp. 19, 23.

Table 5.7 provides a regional breakdown of drinking water quality associated with the water supply system (*vodoprovodnaya set'*). The data suggest that instances of sanitary-chemical pollution are relatively high in the Central, North Western, Ural and Far Eastern federal *okruga*. In contrast, microbiological pollution is most significant in the Southern and Far Eastern federal *okruga*. More generally, the data indicate that in spite of variations in the capacity of cleaning infrastructure and types of economic activity, relatively significant levels of either sanitary-chemical and/or microbiological pollution are registered throughout the Russian Federation.

Table 5.7 Drinking water quality within the water supply system (both surface water and ground water sources), % of tests failing sanitary norms, 2002

Federal *okruga*	Sanitary-chemical tests	Micro-biological tests
Central	21.3	6.8
North Western	24.6	8.2
Southern	11.1	10.7
Volga	16.6	7.5
Ural	27.4	6.8
Siberian	17.3	7.1
Far Eastern	25.4	11.3
Russian Federation average	18.9	8.1

Source: Gossanepidnadzor, 2003, p. 24.

Environmental Issues in Russia's Non-urban Regions

The preceding analysis of air and water pollution trends, while useful in establishing a general understanding of both national and regional patterns of environmental problems, is clearly biased towards Russia's urban regions. This is understandable given the highly urbanised nature of Russian society (73 percent of the total population) and the marked influence of urban areas on ecological systems, which extends far beyond the limits of the built-up area. Nevertheless, such a focus ignores the vast areas of non-urban land, characterised by varied landscape and vegetation types and relatively low population densities (e.g. see Newell, 2004). For example, much of northern European Russia, Siberia and the Russian Far East record population density levels of approximately 3 persons/km^2 (Goskomstat, 1996, p. 717). In contrast, the highest densities of population are found in the southern and central parts of European Russia. For the most part, these are relatively modest and correspond to values of between 40 and 90 persons/km^2. The major exception is the Moscow region (including Moscow city), which registers a figure of more than 300 persons/km². The higher concentration of population in the European parts of Russia is reflected by the fact that more than two-thirds of Russia's population is located west of the Urals (excluding the Ural federal *okrug*) (see Goskomstat, 2002b, pp. 14-16). Dienes (2002) draws attention to the marked contrast between Russia's large urban centres and the significant expanses of intervening non-urban land (see also Ioffe et al., 2004). In recognition of the need to engage more purposefully with the environmental situation beyond Russia's main urban localities, the following section explores the extent of Russia's forest, agricultural and specially protected land resources, and draws attention to some of the main environmental issues.

Russia's Land Resources

In order to gain a feel for the extent of Russia's non-urban regions, it is useful to explore the characteristics of Russia's land resources. These are divided into 7 main administrative categories, which together comprise the land fund (*zemel'nyi fond*). The land fund is a legal categorisation and not necessarily a precise reflection of land type distribution. In addition, there have been marked changes in the size of certain categories during the 1990s, as legal adjustments resulted in the reclassification of certain areas of land (see Table 5.8). The administrative categories comprising Russia's land fund are therefore not immutable and are themselves comprised of a range of different land areas (*ugod'ya*), which can be broadly divided into agricultural areas or farmland (*sel'skokhozyaistvennye ugod'ya*) and a range of non-agricultural areas, including forest/shrubbery and built-up land areas. Despite these caveats, the land fund categories do provide an indication of the relative composition of Russia's total land area.

    Table 5.8 illustrates clearly the limited extent of land designated for urban use and industrial activities in comparison with forest and agricultural land areas. For example, land situated under the jurisdiction of urban and rural administrations (including cities/towns, minor settlements and rural settlements) accounts for just over 1 percent of Russia's total land fund. Furthermore, within this total, land used for agricultural purposes is dominant, with buildings and installations comprising approximately 15 percent of the total, or 3.1 million hectares (IOPRR, 2001, p. 116). The marked absolute increase in the area of this land category during the early-mid 1990s was due largely to the transferral of land from forest and agricultural categories in order to facilitate construction activities (Minpriroda, 1996, p. 38). Land associated with industrial and transport-related activities covers an additional 1 percent of Russia's land fund. As indicated above, these relatively small areas of land receive considerable attention due to the fact that they account for almost three quarters of the total population and are characterised by a marked concentration of polluting activities. For example, in addition to the aforementioned air and water issues, the long-term operation of industrial enterprises and related production units, as well as exhaust emissions from motor vehicles, can generate significant levels of heavy metal soil pollution. This phenomenon is particularly evident in large urban regions such as St. Petersburg and Moscow. Similarly, transport networks can result in the formation of pollution corridors around major urban areas.

Table 5.8 Dynamic of Russia's land fund: Area of main land categories, 1990-2002 (millions hectares)

Land category	1990	1996	2002
Lands for agricultural purposes[a]	639.1	670.1	400.7
Land under the jurisdiction of urban and rural administrations	7.5	38.2	18.9
Land for industry, transport, links etc.	16	18.2	17.2
Specially protected territory	17.4	29.8	34.2
Forest fund[a]	895.5	825.6	1103.1
Water fund	4.1	19.4	27.8
Land reserves[b]	130.2	108.5	107.9
Total	1709.8	1709.8	1709.8

[a] Between 1997 and 1998 a large area of designated agricultural land was seemingly transferred to the forest fund (approx. 200 million hectares). Since 1998, agricultural land has been reported solely under the heading of 'lands for agricultural purposes' (*zemli sel'skokhozyaistvennogo naznacheniya*) in relevant state reports.
[b] Note that the land reserves designation (*zemli zapasa*) refers to 'non-allocated land'.

Sources: compiled from IOPRR, 2001, p. 99; MPR, 2001b, p. 24; MPR, 2003, p. 31.

Agricultural Land

Lands for agricultural purposes (*zemli sel'skokhozyaistvennogo naznacheniya*) accounts for almost one quarter of Russia's total land fund (Table 5.6) and refers predominantly to '...land utilised by agricultural enterprises, organisations and citizens engaged in the production of marketable agricultural output' (IOPRR, 2001, p. 109). It also incorporates land used for agricultural purposes situated within urban areas as well as traditional lands associated with the activities of indigenous peoples. Nevertheless, a significant percentage of this land fund category is comprised of forest in addition to other non-agricultural land-use types. The aforementioned category of farmland (*sel'skokhozyaistvennye ugod'ya*) is a seemingly more useful, and a less ambiguous, indicator, as it incorporates arable, fallow, hay and pasture land, in addition to areas of perennial vegetation, across *all* land categories. Chapter 3 indicated that in 2002 farmland (and this includes farmland located within urban areas) totalled 221 million hectares, or approximately 13 percent of Russia's total land fund. The greatest concentration of farmland is located in the southern parts of European Russia. This is highlighted by the data provided in Table 5.9 with the Central, Southern and Volga federal *okruga* registering approximately 70 percent of arable farmland, and yet accounting for just 14 percent of Russia's total land area.

Table 5.9 Area of arable farmland (*pashnya*) utilised by agricultural enterprises and citizens engaged in agricultural production activities[a]

Federal okruga	Area (mln km²)	Area as a % of Russian total	Arable farmland (000s hectares)	Arable farmland as a % of Russian total	1998 as a % 1990
North Western	1.7	10	3328	2.7	91.9
Central[b]	0.7	4	23746	19.5	93.8
Southern	0.6	4	22078	18.2	95.3
Volga	1	6	37282	30.7	93.7
Ural	1.8	11	8801	7.2	93.0
Siberian	5.1	30	24000	19.7	88.1
Far Eastern	6.2	36	2375	2.0	74.4
Total	17.1	100[c]	121616[c]	100	

[a] Note that these land-users accounted for 88 percent of total farmland and approximately 96 percent of arable farmland in 1998. The data used here are provided by the former *Goskomzem Rossii.*

[b] The Central federal *okrug* incorporates the Black Earth region.

[c] Due to rounding there are slight discrepancies between the column totals and actual total figures.

Source: IOPRR, 2001, pp. 105-107; Kistanov, 2000.

Table 5.10 indicates a modest fall in the area of Russia's arable farmland during the course of the 1990s. It is tempting to blame such trends solely on the problems of soil erosion and related land management concerns highlighted in Chapter 3. Nevertheless, loss of farmland has a long history stretching back into the Soviet period (MEPNR, 1994, p. 41) and, while erosion rates are important, a range of additional factors are also evident (e.g. MPR, 2001a, p. 22). Recent work by Ioffe et al. (2004), concerning farmland abandonment in European Russia, places such trends within the context of Russia's long-term agricultural development (i.e. extensive development during the Soviet period) and the nature of the Russian settlement system. In particular, they argue that:

...the fate of myriad local farms appears to be molded by powerful factors external to agriculture. In the NCZ [non-chernozem zone], land at a distance from a large city is being abandoned, and so the environs of those cities come to look like agricultural oases. In the South, the spatial development is more even, but here, too, excessively arid land is being abandoned and peripheral districts in regions devoid of large urban centers are ailing even on good soils (Ioffe et al., 2004, p. 938).

At the same time, it is evidently difficult to determine a precise figure for this trend of land abandonment due to the inadequacies of related data-sets. Ioffe et al. (2004, pp. 922-923) draw attention to the sluggish response of land area data to the changing situation at the local level. As such, they employ data related to the area of land under crops as a surrogate for the area of farmland, since these figures are produced on a more regular basis. Subsequent data analysis indicates a significant fall (>25 percent) in the area of cropland at the national level during the 1990s, bringing it into line with recorded falls in overall output (see Figure 3.1).

Table 5.10 Changing area of farmland (*sel'skokhozyaistvennye ugod'ya*) across all land use categories (million hectares)

	1986	1991	1993	1995	1999	2001	2003	2003 as a % of 1991
Farmland – total area	228.8	222.4	222.5	221.9	221.1	221.0	220.8	99.3
Including:								
arable land	*134.2*	*132.3*	*132.0*	*130.2*	*125.3*	*123.9*	*122.6*	*92.7*
(% of total)	*(59%)*	*(59%)*	*(59%)*	*(59%)*	*(57%)*	*(56%)*	*(56%)*	

Sources: Goskomekologiya, 2000; IOPRR, 2002a, p. 84; MEPNR, 1995, p. 16; Minpriroda, 1996, p. 37; MPR, 2002, p. 33; MPR, 2004, p. 30.

Thus, while it is clear that soil erosion remains a significant problem for Russia's agricultural regions, the work of Ioffe et al. is suggestive of relatively small (in the context of Russia's total farmland) but, nevertheless, significant areas of farmland reverting to fallow, scrub and forest as prevailing socio-economic conditions encourage land to be abandoned. As such, agricultural activity is withdrawing to core areas, as determined by prevailing environmental conditions and settlement patterns.

Forested Land

Russia is the world's most forested country, accounting for 22 percent of global forest resources (compared to 16 percent for Brazil and 7 percent for Canada [World Bank, 1997, p. 30]). More specifically, the forest fund accounts for approximately two-thirds of Russia's overall land fund and totalled 1103 million hectares in 2002. Table 5.8 indicates the marked increase in the area of the forest fund during the mid to late 1990s, and this was due predominantly to the administrative transferral of forested land parcels from other land-use categories, rather than a marked growth in the actual area of forest (e.g. MPR, 2002, p. 34). The forest fund incorporates significant areas of non-forested land.[7] For example, in 2002, approximately two thirds of the forest fund was classified as 'forested' (*pokrytye lesom*) with the remaining 30 percent comprising water bodies, agricultural areas and reindeer pastures. Significant areas of forested land are also located within the other main land fund categories. In recognition of this, Table 5.11 provides additional data for land area (*ugod'e*) characterised by forest and shrubbery growth calculated across all land categories. This latter figure represents a more accurate estimate of Russia's actual forested area, and accounts for approximately 50 percent of the total land fund area.

Table 5.11 Area of Russia's forest fund and forested land across all land categories (million hectares), 1990-2002

	1990	1993	1995	1997	1999	2002
Forest fund (includes areas of non-forested land)	895.5	843	843	828	1060	1103
Forest and shrubbery growth areas across all land fund categories	-	-	785.6	786.1	-	897.2

- Data missing

Sources: IOPRR, 2001, pp. 99-100; Goskomstat, 1998, p. 29; Minpriroda, 1996, p. 37; MPR, 2003, pp. 31-32.

The composition of Russia's forested land is biased towards coniferous forest (over 70 per cent in 1998) with softwood broadleaf (17 percent) and hardwood broadleaf (less than 3 percent) accounting for relatively small overall percentages. Russia's total estimated reserves of standing timber (*obshchii zapas drevesiny na kornyu*) have remained reasonably steady since the early 1980s, and comprised 81.9 billion m³ in 2001 (Goskomstat, 1998, p. 58; MPR, 2002, p. 51). The largest expanses of Russia's forested land are located in the Siberian and Far Eastern federal *okruga* in addition to northern regions of European Russia (Table 5.12). Indeed, just six federal units, located within these three macro-regions, account for approximately 50 percent of Russia's total forested area (according to gross hectare coverage). These include: Irkutsk *oblast'*, Krasnoyarsk *krai*, the Evenk autonomous *okrug* (all located in the Siberian federal *okrug*), Khabarovsk *krai*, the Republic of Sakha (both Far Eastern federal *okrug*) and the Komi Republic (North Western federal *okrug*). The Republic of Sakha alone accounts for almost 20 percent of the Russian total, although at the same time covering 18 percent of Russia's total land area. The overall dominance of the Siberian and Far

---

[7] The forest fund includes forested land, land earmarked for reforestation and non-forested land considered necessary for the functioning of the forestry sector (IOPRR, 2001, p. 123).

Eastern federal *okruga* is clearly reflected in the available data with these two regions together accounting for two-thirds of Russia's forested land. Data relating the extent of forested land to the area of the federal unit provide more insight into the regional concentration of forested land and highlight the relatively high density of forest growth in northern regions of European Russia, central Siberia and the Russian Far East (Figure 5.6).

Table 5.12   Regional distribution of forested land, 1998[a]

Federal *okruga*	Area (mln km²)	Area as a % of Russian total	Forested land (000s hectares)	Forested land as a % of Russian total
North Western	1.7	10	86934	11
Central	0.7	4	22022	3
Southern	0.6	4	4332	1
Volga	1	6	36683	5
Ural	1.8	11	67199	9
Siberian	5.1	30	269605	35
Far Eastern	6.2	36	283011	37
Total	17.1	100[b]	769785[b]	100[b]

[a] According to GULF (*Gosudarstvennyi uchyot lesnogo fonda*) data, 01/01/98. This accounting procedure was carried out every five years, although the frequency of the procedure appears to have increased in recent years.
[b] Due to rounding there are slight discrepancies between the column totals and the actual total figures.

Source: compiled from IOPRR, 2001, pp. 141-143; Kistanov, 2000.

Problems in the Forestry Sector

Superficially at least, the preceding analysis appears to suggest that the recent period of marked societal upheaval has not undermined significantly the integrity of Russia's forest resources. This observation is reinforced by data concerning forest extraction (*vyvozka*) rates, which fell markedly during the period 1990 to 1998 (Goskomstat, 2000, p. 218) before showing signs of recovery through to 2002 (Goskomstat, 2004a, p. 197). Similarly, official production rates for all major forest products fell during the 1990s, although once again there have been signs of recovery in recent years.[8] For example, timber (*delovaya drevesina*) production declined from 256 million cubic fest-metres in 1990 to just 76 million cubic fest-metres by the end of the decade (Goskomstat, 2000, p. 225). Nevertheless, these official data mask a far more strained situation with regard to Russia's forest resources. First, it is important to acknowledge the existence of weak management systems and wasteful harvesting techniques. Many of these concerns have linkages

---

[8] See Backman and Zausaev (1998) and Backman (1999) for overviews of the forest sector in the Russian Far East and Siberia during the 1990s.

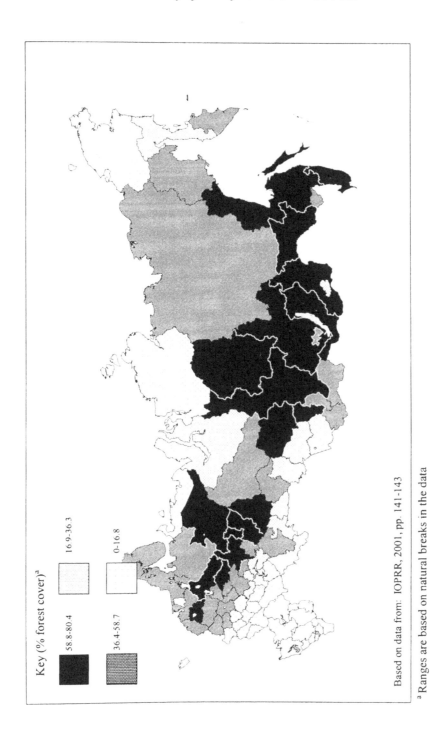

Based on data from: IOPRR, 2001, pp. 141-143

[a] Ranges are based on natural breaks in the data

Figure 5.6   Percentage of land covered by forest, according to federal region (% *lesistosti*)

Key (% forest cover)[a]

58.8-80.4

16.9-36.3

36.4-58.7

0-16.8

with the Soviet era, since forestry harvesting methods during this period were considered inefficient and resulted in the destruction of relatively large areas of forest and a high loss of timber (Goskompriroda, 1990, p. 26; Pryde, 1972; Pryde, 1991, pp. 112-135[9]; Newell and Wilson, 1996; World Bank, 1997). Second, both independent analysts and academics have cast serious doubt on official forest extraction figures in recent years, suggesting that large volumes of timber are being harvested and exported illegally (see Newell, 2004, pp. 70-80). Such activity would appear to be most prevalent in the Russian Far East where state control of the region's immense forest reserves is relatively weak and the economies of China, Japan and South Korea provide a large end market. The old-growth forests along Russia's northwest border with Finland also suffered from unregulated cutting during the 1990s, although the proximity to both Western environmental movements as well as environmental organisations based in Moscow and St. Petersburg facilitated greater levels of monitoring and control activity. Third, there is evidence to suggest that the quality of Russia's forest resources is deteriorating. Newell (2004, pp. 31-32) refers to recent studies highlighting the shrinking areas of frontier forest within Russia (essentially large areas of untouched and continuous forest) and the changing age composition of forested areas, with a noted rise in the proportion of young stands of trees. This latter trend would appear to have its roots in the Soviet period, and is particularly relevant to stands of conifer trees rather than hardwood and softwood broad-leaved stands (see IOPRR, 2001, p. 144). Fourth, the formerly independent Federal Forestry Service was disbanded at the same time as *Goskomekologiya* in 2000 and its management functions transferred to the Ministry for Natural Resources (see Chapter 4 for more on administrative changes). Some observers interpreted this as an attempt by state organs to facilitate the purposeful development of Russia's forest resources over the short- to medium-term. Furthermore, the 1997 Forest Code eased the process of transferring Group I forests (i.e. forests receiving the highest level of protection due to their acknowledged important ecological functions) to one of the other two forest categories, in order to enable further exploitation.[10] Similar concerns have been raised more recently in the light of ongoing legislative revision.

---

[9] Note that this chapter of his book was prepared by Kathleen Braden.

[10] Russia's forest resources are divided into three categories (see World Bank, 1997, p 115). Group I forests are accorded the highest level of protection due to their role in regulating local ecological systems and accounted for 21 percent of the total in 2003 (MPR-LR, 2003, p. 8). Group II forests are limited in extent (approximately 6 percent) and refer to forests with limited commercial potential. Group III forests are designated commercial forests and form the largest category (approximately 70 percent).

(Photograph: author)

Plate 5.2 Bittsevskii nature park, southern fringe of Moscow city

Specially Protected Natural Areas

Russia's specially protected areas fulfil a number of functions related to conservation, scientific and socio-cultural concerns (Ekologiya, 2001, p. 107). According to land fund data, specially protected natural areas accounted for approximately 2 percent of Russia's total land fund in 2002 compared to just 1 per cent in 1990 (IOPRR, 2001, p. 99; MPR, 2003, p. 31). This reflects the transferral of land from other land-use types during the course of the 1990s in order to create various categories of protected land. At the same time, this figure underestimates the actual extent of Russia's specially protected territory of both federal and regional significance (MPR, 2002, p. 34). Thus, the main protected territory designations accounted for approximately 3.5 percent of Russia's total land area in 2002, indicating the sizeable chunks of protected territory residing within the forest fund and other land fund categories (see Minpriroda, 1996, pp. 38-39).

Chapter 2 referred to the origins of the *zapovedniki* network and this grew considerably during the last two decades of the Soviet Union, with 36 such designations created on Russian territory alone (Kostenchuk, 1993, section 4.4). This apparent success should not disguise the marked difficulties suffered by the network of protected territories during the Soviet period (see Weiner, 1999). Nevertheless, Russia inherited a reasonably extensive network in 1991 and further development occurred during the 1990s. A 1995 law 'Concerning Specially Protected Natural Territories' (which built upon the 1991 law Concerning the

Protection of the Natural Environment) laid out the basic structure of Russia's protected areas network (see Table 5.13). It identified seven main categories of specially protected area within the Russian Federation with differing levels of protected status and purpose. These include: *zapovedniki* (*gosudarstvennye prirodnye zapovedniki*), national parks (*natsional'nye parki*), nature parks (*prirodnye parki*), state nature reserves (*gosudarstvennye prirodnye zakazniki*), natural monuments (*pamyatniki prirody*), dendrological parks & botanical gardens (*dendrologicheskie parki i botanicheskie sady*) and health spas and resorts (*lechebno-ozdorovitel'nye mestnosti i kurorty*).

The *zapovedniki* continue to be regarded as the main element within the system of protected natural areas (Pryde, 1997) and, as such, form the basis of Russia's contribution to the world network of Biosphere Reserves. In 2002, *zapovedniki* numbered 100 (68 in 1990) and covered an area of 33.7 million hectares, or 1.5 percent of Russia's total land area (Goskomstat, 2003a, p. 62). National parks have become increasingly important in recent years in terms of overall area. At the beginning of the 1990s there were only 12 national parks covering an area of approximately 1.8 million hectares. By 2002, they numbered 35 and covered approximately 6.9 million hectares, or 0.4 percent of Russia's total land area (Goskomstat, 2003a, p. 62). Many national parks were established between the years 1991 and 1996 (MPR, 2002, pp. 130-131). Other categories of protected natural areas also registered an increase in area during the 1990s.

A significant percentage of Russia's federal regions possess specially protected land in the form of *zapovedniki* or national parks. Unsurprisingly, significant parts of the sparsely populated expanses of Siberia and the Russian Far East have been designated as specially protected territory (Table 5.14). Together, the Siberian and Far Eastern federal *okruga* accounted for approximately 90 percent of Russia's *zapovedniki* territory in 2001. However, it is the North Western and Siberian federal *okruga* which are responsible for a marked chunk of the national park territory. The relative youth of the national park system is reflected in Table 5.14, with significant areas designated after 1991. In contrast, and as noted above, large parts of the *zapovedniki* network were in place before 1991. At the same time, the Ural federal *okrug* benefited from the designation of the Gydanskii *zapovednik* (878 thousand hectares), located in the Yamalo-Nenets autonomous region, during 1996. Designations are easily made on paper, but their overall effectiveness as tools for landscape and biological conservation can be questioned and it is important to put such developments in the wider context. Russia's network of specially protected land is beset by a range of problems and many of these are recognised explicitly in the federal programme 'Support for Specially Protected Natural Areas' which is part of the larger target programme 'Ecology and Natural Resources of Russia, 2002-2010' (see Chapter 4). Not least, funding levels are often inadequate and illegal activities such as poaching are not uncommon, a situation aggravated by the immense size and remoteness of many of the protected sites (e.g. see Pryde, 1997, pp. 68-69). Furthermore, local management of protected areas can be compromised by poor technical infrastructure, limited long-term planning and corrupt activities (e.g. Ekologiya, 2001, pp. 107-108; Graebner, 2001, pp. 15-16; Stepanitsky, 2001, p. 13).

Table 5.13 Main functions of Russia's specially protected natural areas[a] (according to the 1995 law on Specially Protected Areas)

Types of Specially Protected Area	Selected main functions
Zapovedniki	• Protection of natural territory • Preservation of biological diversity • Scientific study of the natural environment, educational role • Preservation of both 'typical' and unique ecological systems • Strict control of activities within the protected territory
National parks	• Nature protection functions in addition to educational, scientific and cultural aims • Regulated tourism function
Nature parks	• Preservation of nature and natural landscapes • Emphasis on recreational activities • Development of measures for the protection of nature within the territory
State nature reserves	• Preservation and restoration functions • Divided into five specific 'type' of reserve: landscape, biological, palaeontological, hydrological and geological
Natural monuments	• Conservation of objects or areas displaying unique or irreplaceable ecological, cultural or aesthetic qualities (includes both natural and artificial objects/areas)
Dendrological parks & Botanical gardens	• Preservation, educational and scientific functions
Health spas and resorts	• Medical treatment and rest functions

[a] A number of smaller and less prestigious specially protected categories also exist. For example, Museum nature preserves (*muzei-zapovedniki*) and Museum estates (*muzei-usad'by*) aim to protect historical and cultural territories with significant natural attributes.

Source: Rossiiskaya Gazeta, March 22, 1995, pp. 9-10.

Table 5.14   *Zapovedniki*[a] and National Parks according to federal *okruga*, 2001

	No. of *zapovedniki*	Area (000s hectares)	% of total area	Area designated *after* 1991 (% of federal *okrug* total)	No. of National Parks	Area (000s hectares)	% of total area	Area designated *after* 1991 (% of federal *okrug* total)
North Western	10	1089[b]	3.5	90 (8)	8	2986	42.9	2213 (74)
Central	11	208[b]	0.7	31 (15)	7	587	8.4	551 (94)
Southern	10	719	2.3	34 (5)	3	349	5.0	55 (16)
Volga	13	626[b]	2.0	85 (14)	8	393	5.6	108 (28)
Ural	7	2431[b]	7.8	878 (36)	3	194	2.8	137 (71)
Siberian	20	14040	45.3	5335 (38)	6	2459	35.3	177 (7)
Far Eastern	24	11856	38.3	4650 (39)	0	-	-	-
Russian Federation	95	30969	100[c]		35	6968	100	

[a] The data refer to those *zapovedniki* placed under the jurisdiction of the Ministry of Natural Resources. The remaining 5 *zapovedniki* are affiliated to the Russian Academy of Sciences (4) and the Ministry for Education (1).

[b] Darvinskii *zapovednik* covers 112 thousand hectares and is divided between the territories of Vologda *oblast'* (North Western federal *okrug*) and Yaroslavl' *oblast'* (Central federal *okrug*). Each federal *okrug* was allotted 56 thousand hectares in the table. Similarly, Yuzhno-Ural'skii *zapovednik* covers 254 thousand hectares and is split between the Republic of Bashkortostan (Volga federal *okrug*) and Chelyabinsk *oblast'* (Ural federal *okrug*). Each federal *okrug* was allotted 127 thousand hectares in the table. The original data refer to 93 *zapovedniki* of the 95 officially under the jurisdiction of the Ministry for Natural Resources at this time. It is possible that the 2 *zapovedniki* incorporating 2 federal territories are counted twice for official accounting purposes and this method has been followed in the table.

[c] Due to rounding there are slight discrepancies between the column totals and the actual total figure.

Source: compiled from MPR, 2002, pp. 127-131.

Concluding Remarks

The marked falls in gross pollution levels across the post-socialist countries of CEE and the FSU have received considerable attention in the Western literature during the last decade or so. However, it is important to appreciate the complex and multifaceted nature of Russia's contemporary environmental situation. Furthermore, current trends are not simply reducible to the ebb and flow of industrial pollution outputs, although such trends are still important. Persistent financial and organisational problems have helped to undermine the consistency and comprehensiveness of environmental monitoring networks, and this has obvious implications for comparative analysis. Bearing these issues in mind, the available data suggest that the quality of the environment improved little during the 1990s, in spite of the significant falls in overall levels of pollution discharge. This is particularly relevant to Russia's urban regions. It is tempting to blame such a situation on the inadequacies of Russia's transformation process. To some extent, this interpretation has validity since Russia's environmental situation would undoubtedly improve if cleaning equipment was upgraded and production processes made more efficient. However, quite apart from the aforementioned particularities of Russia's societal transformation, such a limited approach ignores a range of attendant concerns, which can have a direct or indirect bearing on the current environmental situation.

First, the relationship between gross pollution output and environmental quality is multifaceted and mediated by a range of natural and social factors. As such, an analysis based purely on volume pollution figures is likely to misinterpret key underlying trends. A host of local-level factors have the potential to influence environmental quality, in addition to gross pollution output, and these might include the type and concentration of industrial activity, climatic conditions, the assimilative capacities of local biophysical systems and the nature of local energy sources. Second, it is clear that certain contemporary pollution issues are related directly to the environmental 'baggage' associated with market-type relations. Mention was made of the increasing levels of motor vehicle usage linked to a burgeoning consumer market. Other issues might have been mentioned such as increasing levels of domestic waste production (see Chapter 3) or loss of green land due to housing construction. Third, the strained socio-economic situation within Russia is likely to discourage the emergence of strong environmentalism, as large sectors of society remain preoccupied with household survival. It is clear that environmental concern wields far less influence within government policy circles now than it did at the end of the Soviet period. Some of this evident apathy can be related to the fact that dangerous levels of air and water pollution are often not readily apparent to the inhabitants of urban regions. As with millions of Westerners living in the large urban conurbations of London, New York or Los Angeles, many Russians are faced with levels of air pollution which exceed certain sanitary limits, but the health effects of such pollutants are not perceptible unless observed systematically over the long-term. This allows environmental concerns to be sidelined by government departments, industrial enterprises and individuals in the face of more urgent daily concerns. Fourth, the pervasiveness of the Soviet legacy should not be underestimated. The influence of certain types of environmental pollution, such as that associated with the Chernobyl' explosion, on public health and biophysical systems is difficult to determine with accuracy. In addition, the Soviet legacy can also have indirect consequences for the contemporary environmental situation. For example, the economic malaise characteristic of the late Soviet period ensured that vast parts of the technostructure (municipal infrastructure, industrial equipment, transport links, oil and gas pipelines etc.) bequeathed to the Russian Federation were in urgent need of investment and upgrade (see Chapter 3). This would present a significant task to any country, quite apart from one that subsequently embarked on a further decade of societal restructuring and transformation. The enormity of this task is possibly a main reason for Russia's apparent tendency to resort to reactive, rather than proactive, stances in addressing environmental problems.

The regionally varied nature of Russia's environmental situation formed an integral focus of the chapter. It is clear that certain regions of Russia are subject to more extreme levels of pollution, or else specific types of environmental problem, and that these are due to the interaction of various social and natural factors over a given period. Air and water pollution issues are particularly acute in the population centres of south-eastern and central Siberia, the southern Urals region as well as central European Russia. Furthermore, there has been limited change in the geography of these centres of pollution during the period 1991 to 2003. At the same time, the geography of Russia's settlement pattern and economic activity ensures that vast tracts of Russia remain relatively untouched by the polluting activities of industrial production, and this is particularly relevant to the northern regions of European Russia, Siberia and the Russian Far East. In these regions, pollution issues are typically replaced by concerns over the rate of natural resource appropriation arising from weak administration, limited state financing and the interplay of domestic and international actors intent on resource extraction and utilisation.

Chapter 6

Concluding Remarks

Introduction

The findings of this study are indicative of the complex and dynamic nature of Russia's contemporary environmental situation. Importantly, they suggest that current trends cannot be wholly explained by reference to falling pollution levels or generalised stereotypes of environmental crisis. Furthermore, the study highlights the marked regional variation in the character and extent of environmental issues across the expanse of the Russian Federation. While such recognition is not particularly novel, it is common for work dealing with Russia's contemporary environmental situation to remain at a generalised level. Where regional case studies are developed (e.g. Crotty, 2003; Holm-Hansen, 2002), it is often difficult to locate the relevance of such work within a national context. The evident deficiencies of both nationally- and regionally-focussed studies serve to emphasise the paucity of understanding related to the regional variation of environmental issues within Russia. This situation is further aggravated by the limited nature of certain data-sets, which provide a restricted insight into key socio-economic and environmental trends. Bearing such concerns in mind, there is much to gain from a historically-sensitive engagement with the varied constellations of societal factors and assorted capacities of biophysical systems that underscore regional variation in environmental concerns. As indicated in the introduction, such an approach is beyond the scope of this particular study. However, its more modest engagement with the changing relationship between society and the wider environment since 1991 provides a necessary insight into Russia's contemporary environmental situation.

The Relevance of the Past

The influence of the Soviet and, indeed, pre-Soviet experience on Russia's contemporary environmental situation is difficult to determine with any accuracy. There is the constant danger that certain elements will be overemphasised while others are underplayed or overlooked. Nevertheless, accepting that the past has a role to play in determining the nature of Russia's current environmental situation is an important starting point. For example, the crisis environmental zones identified towards the end of the 1980s across the territory of the former Soviet Union drew attention to a range of environmental pressures that continue to exert a major influence on natural environmental systems and human health at the local level and beyond. At the same time, they represent but one element of the contemporary environmental situation, and must be placed within the wider context of regionally varied societal change (e.g. due to the differing capacities of local government and municipal authorities to mitigate problems, variation in the ability of regions to attract new economic activity and their proximity to overseas markets etc.) and related to the resilience and recuperative capabilities of biophysical systems at various scales.

The selective way in which past patterns and processes are often incorporated into contemporary explanatory frameworks was highlighted in Chapters 1 and 2. Here it was noted that while the pollution legacies of the Soviet development model were easily imported into post-1991 interpretations of Russia's environmental situation, other legacies relating to the environmental capacities of the Soviet system were largely ignored. This reflects a number of underlying preconceptions and biases. First, the marked extent of the Soviet Union's environmental deficiencies was considered in many quarters as further proof of the failure of the socialist project. In addition, it was relatively easy to posit improving environmental trends as part of the wider triumphalist imposition of liberal democratic ideals across the region. Second, the tendency to posit the break-up of the Soviet Union in 1991 as a fundamental temporal rupture discouraged engagement with the mores and values of the Soviet period as potential path-shaping elements. Third, and following on from the previous point, the covert nature of much 'environmental' activity during the Soviet period ensured that there was little appreciation in the West of the potential resources embedded in Russian society capable of addressing environmental issues. Revisionist work of any kind was further deterred by the sensitive and sometimes horrific nature of Soviet history. Under such conditions, it was natural that many scholars shied away, at least initially, from the delicate task of exposing underlying subtlety and nuance. The overwhelming presence of the Soviet Union in the Western imagination also limited interrogation of environmental thought and insight (in its widest sense) traceable to Russia's pre-revolutionary period. While not wanting to defend the environmental record of the Soviet Union, which was by any standard extremely poor, it has been a main contention of this study that a more sensitive understanding of Russia's environmental legacies is essential if we are to move away from the notion of Soviet society possessing limited social and intellectual capital in the general area of environmental protection. For example, the connections between current Russian understandings of sustainable development and pre-revolutionary thought were highlighted in the exploration of recent policy and legislative change (see Chapter 4). Similarly, Russia's contemporary environmental movement is cast in a new light once Soviet and pre-revolutionary precursors are acknowledged. The Soviet regime also left a reasonably extensive environmental monitoring infrastructure in addition to systems of natural resource management and environmental impact assessment. The protected land system, centred on the *zapovedniki* network, provides a further example of the continuities between the pre-Soviet, Soviet and post-Soviet periods. An appreciation of these varied environmental sensibilities and capacities embedded deep in Russian society has implications for the nature of international assistance and guidance. In particular, such understanding highlights the inadequacies of those models of cooperation and assistance based on the wholesale imposition of Western procedures and institutional structures.

The Western Development Model and the Environment

Just as Russia's contemporary environmental situation cannot be explained wholly by reference to past pollutions, at the same time, notions of marked environmental improvement associated with the implementation of market-type infrastructure or the adoption of democratic practices should also be treated with caution. This is not to dismiss out of hand the environmental sensibilities of the Western development model. In a restricted sense, elements of this model have much to offer the Russian Federation in helping to alleviate existing environmental problems over the short- to medium-term via the importation of new technology and management frameworks. Indeed, the Russian economy in the early 1990s had considerable scope for reducing its polluting potential due to the weakening of central control and simultaneous exposure to international forces of demand and supply. Furthermore, international lending agencies and foreign investors provided, and continue to provide, conduits for good-practice and technical innovation in certain instances. Democratic systems of governance promise additional benefits associated with the increased transparency and accountability of state activity. A further and more general point worthy of consideration concerns the global dominance of capitalist values and associated institutional structures. This situation reinforces the need to relate the Russian environmental experience to the rhythms and nuances of capitalistic processes, at least for the foreseeable future. The alternative, it would seem, is for critical analysis to degenerate into either doom-laden predictions of environmental crisis or else utopian visions of future nature-society relations with limited opportunity for successful realisation.

Despite the positive elements associated with the implantation of the Western development model, it is still necessary to acknowledge the particularities of Russia's development path and accept that these are unlikely to coincide with the characteristics of a perfectly functioning market-based democracy. The experiences of the Czech Republic, Poland and other countries of CEE have tended to support crude transition models of environmental change, with their relatively successful movement to a market economy being accompanied by falling pollution discharges and improving levels of economic efficiency. Much has been made of the apparent 'decoupling' of economic growth trends from pollution emissions across central and eastern Europe in recent years. The expectation that Russia will follow a similar trajectory ignores the underlying differences extant within the former socialist countries of CEE and the FSU. These encompass not only contemporary factors, such as the marked financial assistance associated with the EU accession process, but also the varied experiences of socialism evident throughout the region. Indeed, much of CEE was characterised by a relatively brief engagement with socialist economic practice in comparison with Russia. Chapter 3 drew attention to the persistence of relatively high levels of pollution intensity and correspondingly low levels of resource efficiency within the Russian economy during the 1990s, and this in spite of marked falls in gross pollution levels. Furthermore, the noted decoupling of economic and pollution indicators evident in countries such as the Czech Republic is not so clear cut in the Russian Federation. While levels of polluted drainage discharge continue to exhibit a slight downward trend despite Russia's improving economic performance, stationary source air pollution emissions rose by approximately 1.5 million tonnes over the period 1999-2003. At the same time, it was noted in Chapter 3 that the data for polluted drainage discharge may in fact underestimate actual output levels due to changes in statistical accounting procedures. It is tempting to blame the absence of strong decoupling trends on the inadequacy of Russia's economic transition exemplified by the 'deviant' interaction of economic and political factors in key sectors such as natural resource extraction and processing. Assertions of cronyism combined with the associated activities of the oligarchs during the 1990s reinforce such interpretations. However, this approach fails to engage with the historical particularities of Russian society grounded, for example, in the political-economic relationships forged during seven decades of Soviet rule, and the likelihood that such 'deviations' will persist into the future. More generally, this type of conclusion ignores the possibility that difference will be maintained and reproduced over the medium-term as more resistant elements of Russia's socialist past recombine with contemporary patterns and processes.

While an exploration of the specificities of Russia's development path is a necessary task in order to understand more fully contemporary environmental pressures, there is also much to gain from placing Russia's experience in a broader analytical framework. Thus, while the movement towards a market-type system promises to improve levels of resource use efficiency and reduce pollution output within certain sectors of the Russian economy, the long-term implications of this restructuring process for regional, national and global environments are far from certain. It is within such a long-term framework that a fundamental environmental critique of Russia's market-led restructuring process has most salience. This reflects the fact that mature market economies are characterised by a range of distinct environmental issues with global ramifications. For example, reports focussing on environmental concerns in the European Union highlight the persistent problem of urban-based air pollution (principally with respect to concentrations of low-level ozone and fine particulates) as well as increasing levels of domestic waste production and energy consumption (EEA, 2004).

The Contemporary State of the Russian Environment

As a means of moving beyond limited understandings of the contemporary Russian environment associated with notions of crisis or environmental improvement, the study engaged with the specifics of Russia's environmental situation at both the national and the sub-national level. What emerges is a picture of marked regional variation in both the nature and extent of environmental pressure related strongly to Soviet and pre-revolutionary patterns of economic expansion. The regional picture is difficult to discuss in general terms. Dienes' articulation of the 'Russian archipelago' provides a starting point for assimilating the 'islands' of environmental stress (i.e. main urban centres) and the large expanses of land characterised by limited economic activity within a broader analytical framework (see also Ioffe et al., 2004). Such an approach is particularly helpful in preventing an overemphasis on urban-centred issues. At the same time, the nature of environmental concern can vary markedly between urban regions, and tends to reflect differences in prevailing economic activities. Certain urban areas are benefiting from relatively high levels of economic connectivity, at both the domestic and international scale, due to the particularities of previous development patterns, or else opportunities presented by the changing geopolitical situation and Russia's increased openness to global processes. Conversely, other urban regions, and particularly those located at a distance from large urban centres, are characterised by limited opportunities to engage in meaningful economic relationships due to their relative inaccessibility. The absence of simple multiplier effects is further compounded in some regions by the specificities of Russia's inherited industrial geography. For example, Siberia and the Russian Far East have traditionally served as resource hinterlands for the core regions of European Russia (e.g. see Rodgers, 1990). As a consequence, urban economies within these regions are relatively underdeveloped and lack the necessary infrastructure (e.g. transport, energy etc.) to benefit from their wealth of natural resources (e.g. see Dienes, 2002; Shaw, 1999, 94-114).

Russia's regional economic situation must also be related to the policies of central government. The fiscal tussle between the centre and the federal regions was a key focus for Western economic observers during much of the 1990s and remains a point of contention, with regional economies eager to gain greater autonomy over the spending of their tax returns (e.g. see Bradshaw and Vartapetov, 2003, p. 404). Furthermore, Putin's centralising policies, epitomised by the establishment of seven Presidential *okruga* in 2000, continue to undermine the ability of certain regions to take full advantage of rents from natural resource extraction. The influence of global economic processes further complicates the regional picture. The combined actions of international lending agencies and big business can often result in ambiguous outcomes for the local and regional economy. For example, the import of 'know-how' and technology, or the provision of development loans, can help reduce localised pollution output and improve resource use efficiency in a given enterprise or locality, while simultaneously engendering social dislocation and the weakening of existing nature-society linkages in contiguous areas. More specifically, the operations of large multinational corporations, intent on resource appropriation, can result in limited benefit for the local community and even undermine the sustainability of regional economies and ways of life. As such, there would seem much to gain from thinking critically about the ways in which global economic flows and processes can be incorporated more effectively with existing socio-economic and political structures operating at the local level. At present, while the many actors implicated in development processes at the sub-national scale within the Russian Federation (e.g. government agencies, local community representatives, international lending agencies, global business actors etc.) might each claim adherence to the core tenets of sustainable development, there is limited conceptual understanding, or indeed practical experience, to ensure an effective synthesis of their different activities.[1]

Environmental Issues in Russia's Urban Regions

In spite of the marked falls in levels of pollution output during the 1990s, Russia's air and water pollution discharges remain significant at the global level. Furthermore, as noted above, the upturn in economic growth since the late 1990s has prompted a similar increase in stationary source air emissions, in contrast to the experience of many former socialist countries in CEE. Indeed, the improved performance of industrial sectors, such as metallurgy and chemical production, promises to generate additional problems in a range of urban areas located in the middle Volga region and southern Siberia. Figure 5.3 highlighted those urban regions characterised by relatively severe air quality problems during the early part of the twenty-first century. Distinct urban clusters are located in the southern industrialised regions of Siberia (encompassing Irkutsk *oblast'*, Krasnoyarsk *krai* and Kemerovo *oblast'*), the southern Urals region (e.g. Sverdlovsk *oblast'* and Chelyabinsk *oblast'*) and the west Siberian oil and gas producing areas. Additional air quality problems are evident in urban areas located within southern European Russia (e.g. urban regions of Rostov-na-Donu, Krasnodar) and the Russian Far East (e.g. urban regions of Yuzhno-Sakhalinsk, Khabarovsk). It would appear that the noted rise in motor vehicle numbers is encouraging a change in the composition of pollutants within large urban regions, as levels of nitrogen oxides, carbon monoxide and particulate emissions attain greater significance. Turning to water pollution issues, it is clear that water quality remains a key area of concern throughout Russia. Nevertheless, the substantial levels of polluted drainage discharge attributable to large urban areas such as Moscow and St. Petersburg, in addition to the southern regions of European Russia and the North Caucasus, reflect the importance of municipal and agricultural discharges for this particular form of pollution.

In common with other post-socialist countries, the structural changes taking place within Russia's economy have encouraged a range of additional urban-based environmental pressures related to the logic of market-type economies. Increases in waste production and unregulated housing construction pose problems for many urban regions. Importantly, a combination of factors undermines effective engagement with these types of problem, even when policy initiatives indicate awareness of such issues. Weaknesses include limited

[1] The author would like to thank participants at the ESRC-funded seminar 'Sustainable Community Development, Social Impact Assessment and Anthropological Expert Review', held at SPRI, University of Cambridge, 26/11/2004, for useful insight and discussion concerning these issues relative to the Sakhalin region of Russia. This seminar formed part of the ESRC seminar series: 'Trans-sectoral Partnerships, Sustainability Research and the Oil and Gas Industry in Russia', see: http://www.spri.cam.ac.uk/events/russianoil/.

regulatory capacities, inadequate capital expenditure by government agencies as well as deteriorating technical infrastructure (both industrial and municipal).

Environmental Issues in Russia's Non-urban Regions

The traditional focus on air and water quality in urban regions ensures that environmental trends across vast expanses of Russia are little understood. This makes little sense, since many of Russia's most acute environmental problems in the contemporary period are found in non-urban areas. Localised instances of air and water pollution are still apparent in regions with low levels of urban population. For example, agricultural activity continues to generate significant volumes of water pollution in parts of southern European Russia and elsewhere. At the same time, agricultural land forms just one element of the Russian landscape with areas of steppe, forest and tundra dominating substantial parts of European Russia, Siberia and the Russian Far East. General engagement with these regions is limited and often fails to reflect the underlying dynamism of the contemporary situation. For example, with respect to agricultural land, Chapter 5 drew attention to trends of land abandonment in marginalised agricultural areas in response to the changing economic situation during the 1990s. More generally, it is in the spaces between and beyond Russia's main urban regions that weak state regulation and economic activity (both domestic and foreign) are combining most forcefully to generate problems related to resource appropriation and effective land management. A main area of concern is the unregulated harvesting of timber characterising sizeable parts of Russia's forest reserves. This trend is particularly evident in parts of the Russian Far East and southern Siberia. In addition, structural changes within Russia's fishing industry have prompted unsustainable modes of activity in certain instances, as fishing concerns refocus on high-value marine species and export markets in order to achieve the greatest profit (see Nilssen and Hønneland, 2001). More generally, illegal practices in relation to natural resource appropriation, as well as poaching activities, are evident within many parts of Russia, including areas designated as specially protected territory. At one level, the relatively extensive network of protected natural territory provides a potential basis for protecting key landscape elements and wildlife species. However, the limited availability of funding, coupled with the considerable extent of many protected areas, undermines the effectiveness of the available monitoring and regulatory activities within such regions.

Governing the Russian Environment

The environmental problems evident in Russia's non-urban regions are suggestive of the inadequacies of government regulation at all levels, as state agencies struggle to negotiate the competing needs and pressures of local populations and international economic actors within the context of deep-seated societal change. The uncertainties generated by such a situation provide the opportunity for corrupt activity to flourish, as evidenced by unregulated forestry activity in the Russian Far

East. While environmental sensibilities are embedded in the country's 1993 Constitution and reflected in a range of key laws, the weak position of environmental concern within the Russian government has intensified during recent years in deference to the harsh realities of societal change. Furthermore, the economic bias of internationally-sponsored restructuring initiatives and loan deals has often encouraged issues of environmental management and regulation to be cast aside, with the assumption that they will be dealt with later once the Russian economy is functioning 'correctly'. Chapter 4 outlined Russia's engagement with the concept of sustainable development and indicated the gap between policy rhetoric and action. Russia's persistent failure to advance a national strategy for sustainable development is symbolic of the rather limited effectiveness of recent policy change in this area. More generally, the weakness of environmental policy is accentuated by the continuous restructuring of Russia's environmental administration as well as the restricted availability of financial resources. Some commentators have referred to the gradual downgrading of environmental functions within the administrative infrastructure as a process of 'de-ecologisation', culminating in the abolition of *Goskomekologiya* in 2000. The increasingly utilitarian activities of the Russian government under Putin in areas of natural resource appropriation, epitomised by the growing dominance of the Ministry for Natural Resources, is suggestive of a relatively limited concern for addressing environmental issues over the short- to medium-term.

Russia's engagement with international environmental discourse and debate is also worthy of consideration. At one level, Russia displays an appetite for addressing global environmental issues highlighted by its support for a range of treaties and accords during the course of the 1990s, as well as recent initiatives such as its hosting of the World Conference on Climate Change in 2003 (see also Ichikawa et al., 2002; MPR, 2003, pp. 332-340). Russia has been active in emphasising its importance for the integrity of global biophysical systems and this theme has been a recurring feature of both academic debate and state rhetoric (e.g. Oldfield et al., 2003). Such arguments dovetail with Western fears over the environmental threat posed by Russia in the contemporary period (e.g. see Darst, 2001). In short, there would appear to be a gap between the positive environmental image projected on the international stage at gatherings such as the 2002 World Summit for Sustainable Development in Johannesburg, and the corresponding effectiveness of prevailing domestic policies. Russia's self-interest is understandable and mirrored in the actions of Western nations, with the USA providing an obvious example. Nevertheless, there is a clear need for more analytical work in this area in order to explore the medium- to long-term implications of current policy directions.

Russia's protracted negotiations concerning the ratification of the Kyoto Protocol draw attention to the highly politicised nature of international environmental negotiation. The specific reasons behind the initial hesitancy of the Russian regime are difficult to isolate. Economic concerns were clearly important, with fears raised over the viability of an international carbon market following the withdrawal of the USA from the Protocol in 2001. Furthermore, a main argument of Putin's economic advisor, Andrei Illarionov, during 2003-2004 related to the limited potential for Russia to maintain its current carbon surplus in the light of improved domestic economic performance. Ultimately, wider political concerns appear to have played a part in the final decision to ratify, with Russia hoping to gain EU support for its membership of the World Trade Organisation in return for its backing of the Protocol.[2] It is easy to dwell on the apparent inadequacies of Russia's activity in relation to both national and global environmental concerns. However, this would ignore the immense difficulties faced by Russia as it attempts to wrestle with the competing concerns of economic and environmental stability. In truth, effective role models are difficult to come by. More generally, it is perhaps too easy to be critical of Russia's recent performance concerning its development and enforcement of environmental policy. Its failure to position environmental concerns, or indeed other welfare needs, at the centre of the restructuring process is not peculiar to Russia but remains, arguably, a defining feature of global capitalism (e.g. Martin, 2001, p. 191).

Ways Forward

In conclusion, it is appropriate to say a few words concerning the possible future directions of work in the area of Russian society-environment relations. First and foremost, it is clear that the general conceptual frameworks that have dominated Western understanding of Russia's environmental situation require development. In particular, it is important to resist generalisations and engage meaningfully with the particularities of environmental change across the country. Russia's contemporary environmental issues are spatially varied and a consequence of the interplay of old and new processes operating at multiple scales. At the same time, this should not disguise the value of exploring trends common to the country as a whole, since Russia remains an important actor for the effective alleviation of global environmental pressures.

Second, the move to the market should not be used as an excuse to ignore ongoing environmental concerns in the short- to medium-term. The establishment of capitalistic relations will not result in a *de facto* improvement in the regional, national or, indeed, global environmental situation. The imposition of market structures is redefining the way in which economic and social activity takes place across the Russian Federation and, while certain environmental pressures are reduced, others are exacerbated and new types of pressure emerge. Moreover, concerns such as waste production and energy use remain critical issues for the developed economies of the West. In recognition of Russia's current development path, there would appear to be little hope for the moderation of such key environmental issues in the near future. Furthermore, the global dominance of capitalism encourages a critical assessment of the ways in which environmental pressures can be successfully negotiated within the context of prevailing socio-economic structures. This demands, amongst other things, the capacity for critical

---

[2] For example, see 'Russia backs Kyoto climate treaty', BBC News, http://news.bbc.co.uk/2/hi/europe/3702640.stm (accessed January 2005).

reflection and evaluation at all scales from the local through to the global level. Recent initiatives, under the guise of sustainable development, represent an attempt to incorporate environmental concerns with other welfare issues. Furthermore, the holistic pretensions of the concept, while in practice often riddled with contradiction, at least point towards the required scope for any meaningful reworking of the existing state of affairs. Mechanisms for assessing effectively the consequence of the cumulative activities of local, national and supranational actors within a given locality are difficult to envisage. Nevertheless, there would seem much to gain from empowering Russia's environmental governance structures at both the national and regional level, whilst simultaneously ensuring greater reflexivity within the organisational infrastructure of big business and international financial agencies operating within the country. Ultimately, Russia's importance for the functioning of global biophysical systems, as well as the success of global environmental action, strengthens the argument for achieving a more comprehensive understanding of contemporary environmental processes within the country.

Appendix

Interviews and discussions were conducted with representatives from the following organisations and institutions during fieldwork in the Russian Federation (1997-2004):

- Centre for the Preparation and Implementation of International Projects on Technical Assistance, Moscow
- Centre for Russian Environmental Policy, Moscow
- Delegation of the European Commission in Moscow, Moscow.
- Faculty of Geography, Moscow State University
- Greenpeace Russia, Moscow
- Geopolis Consulting Engineers, Moscow
- Institute of Geography, Russian Academy of Sciences, Moscow
- Institute of Global Climate and Ecology, Moscow
- Institute of Sociology, Russian Academy of Sciences, Moscow
- Moscow office of the United Nations Development Programme (UNDP)
- Moscow office of the United Nations Educational, Scientific and Cultural Organisation (UNESCO)
- Moscow State Committee for the Environment and Natural Resources, Moscow
- Russian Ecological Federal Information Agency (REFIA), Moscow
- State Committee of the Russian Federation for Environmental Protection, Moscow

Bibliography

Åhlander, A.S. (1994), *Environmental problems in the shortage economy: The legacy of Soviet environmental policy*, Edward Elgar, Aldershot.

Amirkhanov, A.M. (1997), *The First National Report of the Russian Federation: Biodiversity conservation in Russia*, State Committee for Environmental Protection, Moscow.

Andonova, L.B. (2004), *Transnational politics of the environment: The European Union and environmental policy in central and eastern Europe*, The MIT Press, Cambridge (MA).

Backman, C.A. (1999), 'The Siberian forest sector: Challenges and prospects', *Post-Soviet Geography and Economics*, Vol. 40(6), pp. 453-469.

Backman, C.A. and Zausaev, V.K. (1998), 'The forest sector of the Russian Far East', *Post-Soviet Geography and Economics*, Vol. 39(1), pp. 45-62.

Bailes, K.E (1990), *Science and Russian culture in an age of revolutions: V.I. Vernadsky and his scientific school, 1863-1945*, Indiana University Press, Indianapolis (IN).

Bater, J.H. (1989), *The Soviet Scene: A geographical perspective*, Edward Arnold, London.

Beck, U. (1996), 'Risk society and the provident state', in S. Lash, B. Szerszynski and B. Wynne (eds.), *Risk, Environment & Modernity*, Sage Publications, London, pp. 27-43.

Bityukova, V.R. and Argenbright, R. (2002), 'Environmental Pollution in Moscow: A micro-level analysis', *Eurasian Geography and Economics*, Vol. 43(3), pp. 197-215.

Bøhmer, N. (1999), *Measures for securing radioactive waste and spent nuclear fuel in Murmansk and Archangelsk regions* (Bellona Working Paper), The Bellona Foundation, Oslo (http://www.bellona.no/en/international/russia/waste-mngment/wp_1-1999/index.html, accessed July, 2004).

Bøhmer, N. and Nilsen, T. (1995), *Reprocessing plants in Siberia* (Bellona Working Paper 4), The Bellona Foundation, Oslo (http://www.bellona.no/imaker?sub=1&id=9173, accessed July, 2004).

Bond, A.R., Barr, B.M., Braden, K.E. et al. (1990), 'Panel on the state of the Soviet environment at the start of the nineties', *Soviet Geography*, Vol. 31(6), pp. 401-468.

Bond, A. and Sagers, M.J. (1992), 'Some observations on the Russian Federation Environmental Protection Law', *Post-Soviet Geography*, Vol. 33(7), pp. 463-474.

Bradshaw, M.J. (2002), 'The Changing Geography of Foreign Investment in the Russian Federation,' *Russian Economics Trends*, Vol. 11(1), pp. 33-41.

Bradshaw, M.J. and Bond, A. (2004), 'Crisis amid plenty revisited: Comments on the problematic potential of Russian oil', *Eurasian Geography and Economics*, Vol. 45(5), pp. 352-358.

Bradshaw, M.J.B. and Hanson, P. (1998), 'Understanding regional patterns of economic change in Russia: an introduction', *Communist Economies and Economic Transformation*, Vol. 10(3), pp. 285-304.

Bradshaw, M.J.B. and Stenning, A. (2004), 'Introduction: transformation and development', in M.J.B. Bradshaw and A. Stenning (eds.), *East central Europe and the former Soviet Union: The post-socialist states*, Pearson, Harlow (Essex), pp. 1-32.

Bradshaw, M.J.B. and Swain, A. (2004), 'Foreign investment and regional development', in M.J.B. Bradshaw and A. Stenning (eds.), *East central Europe and the former Soviet Union: The post-socialist states*, Pearson, Harlow (Essex), pp. 59-86.

Bradshaw, M.J.B. and Vartapetov, K. (2003) 'A new perspective on regional inequalities in Russia', *Eurasian Geography and Economics*, Vol. 44(6), pp. 403-429.

Bridges, O. (1992), 'A Comparison in Water Quality: the UK and the CIS', *The Environmentalist*, Vol. 12(4), pp. 255-260.

Bridges, O. and Bridges, J.W. (1995), 'Comparison of air quality in the UK and Russia', *The Environmentalist*, Vol. 15, pp. 139-146.

Bridges and Bridges (1996), *Losing hope: The environment and health in Russia*, Avebury Press, Aldershot.

Burawoy, M. and Verdery, K. (eds.) (1999), *Uncertain Transitions: Ethnographies of change in the postsocialist world*, Rowman & Littlefield Publishers, Lanham (Maryland).

Caldwell, L.K. (1984), *International Environmental Policy: Emergence and Dimensions*, Duke Press Policy Studies, Duke University Press, Durham/North California.

CIS STAT (1996), *Okruzhayushchaya sreda v Sodruzhestve nezavisimykh gosudarstv: Statisticheskii sbornik*, Mezhgosudarstvennyi statisticheskii komitet sodruzhestva nezavisimykh gosudarstv, Moscow.

Clark, E. and Soulsby, A. (1999), *Organizational change in post-communist Europe: Management and transformation in the Czech Republic*, Routledge, London.

Constitution (1985), *Constitution (Fundamental Law) of the Union of Soviet Socialist Republics (adopted 1977)*, Novosti Press Agency Publishing House, Moscow.

Crate, S. (2002), 'Co-option in Siberia: The case of diamonds and the Vilyuy Sakha', *Polar Geography*, Vol. 26(4), pp. 418-435.

Crotty, J. (2002), Economic transition and pollution control in the Russian Federation: Beyond pollution intensification? *Europe-Asia Studies*, Vol. 54(2), pp. 299-316.

Crotty, J. (2003), 'The reorganization of Russia's environmental bureaucracy: Regional responses to federal changes', *Eurasian Geography and Economics*, Vol. 44(6), pp. 462-475.

Dalton, R.J., Garb, P., Lovrich, N.P. et al. (1999), *Critical Masses: Citizens, nuclear weapons production, and environmental destruction in the United States and Russia*, The MIT Press, Cambridge (MA).

Danilov-Danil'yan, V.I. and Yablokov, A. (1999), 'S"ezd zayavil o deekologizatsii gosudarstvennogo upravleniya v Rossii', *Zelenyi mir*, No.14, pp. 8-9.

Darst, R.G. (2001), *Smokestack Diplomacy: Cooperation and conflict in East-West environmental politics*, The MIT Press, Cambridge (MA)/London.

DeBardeleben, J. (1985), *The environment and Marxism-Leninism: The Soviet and East German experience*, Westview Press, Boulder (CO).

DeBardeleben, J. (1990), 'Economic reform and environmental protection in the USSR', *Soviet Geography*, Vol. 31, No. 4, pp. 237-256.

DeBardeleben, J. and Galkin, A.A. (1997), 'Electoral behaviour and attitudes in Russia: do regions make a difference or do regions just differ?', in P.J. Stavrakis, J. DeBardeleben, and L. Black (eds.), *Beyond the monolith: the emergence of regionalism in post-Soviet Russia*, The Woodrow Wilson Center Press, Washington, D.C. and The Johns Hopkins University Press, Baltimore (MA), pp. 57-80.

Deklaratsiya (2003), '"Zelenoe" dvizhenie i grazhdanskoe obshchestvo', *Zelenyi mir*, No. 21-22, p. 3.

Dienes, L. (2002), 'Reflections on a geographic dichotomy: Archipelago Russia', *Eurasian Geography and Economics*, Vol. 43, No. 6, pp. 443-458.

Dienes, L. (2004), 'Observations on the problematic potential of Russian oil and the complexities of Siberia', *Eurasian Geography and Economics*, Vol. 45, No. 5, pp. 319-345.

Dryzek, J.S. (1997), *The Politics of the Earth: Environmental discourses*, Oxford University Press, Oxford.

Dumnov, A.D. (2002), Statistika okruzhayushchei prirodnoi sredy: genezis, predmet i zadachi izucheniya, informatsionno-analitichsekii apparat', *Ispol'zovanie i okhrana prirodnykh resursov v Rossii*, No. 3, pp. 36-62.

EEA (2003), *Europe's environment: The third assessment (Environmental assessment report No. 10)*, EEA, Copenhagen.

EEA (2004), *EEA Signals 2004: A European Environment Agency update on selected issues*, EEA, Copenhagen.

EBRD (2001), *Transition Report 2001: Energy in transition*, EBRD: London

Ekologiya (2001), *Federal'naya tselevaya programma "Ekologiya i prirodnye resursy Rossii (2002-2010 gody)"*, MPR, Moscow.

Energy Strategy (2003), *Energeticheskaya strategiya Rossii na period do 2020 goda*, Approved by ruling of the Russian government Regulation, 28/08/2003, No. 1234-r, Moscow.

EPA (1992), 'The Guardian: Origins of the EPA', *EPA Historical Publication*, No. 1 (Spring) (www.epa.gov/history/publications/index.htm, accessed November, 2004).

Escobar, A. (1995), *Encountering development: The making and unmaking of the Third World*, Princeton University Press, Princeton (NJ).

European Commission (2001a), *Communication from the Commission: EU-Russia Environmental Co-operation*, COM(2001) 772 final, European Commission, Brussels.

European Commission (2001b), *Green Paper: Towards a European strategy for the security of energy supply*, European Commission, Luxembourg.

European Commission (2003), *National Indicative Programme: Russian Federation (2004-2006)*, European Commission, Brussels.

Fagan, A. and Jehlička, P. (2003), 'Contours of the Czech environmental movement: A comparative analysis of Hnuti Duha (Rainbow Movement) and Jihoceske matky (South Bohemian Mothers), *Environmental Politics*, Vol. 12, No. 2, pp. 49-70.

Fagin, A. (2001), 'Environmental capacity building in the Czech Republic', *Environment and Planning A*, Vol. 33, pp. 589-606.

Feshbach, M. and Friendly, A. (1992), *Ecocide in the USSR: Health and nature under siege*, Basic Books, New York.

Fitzpatrick, S. (1999), *Everyday Stalinism. Ordinary lives in extraordinary times: Soviet Russia in the 1930s*, Oxford University Press, Oxford.

Flynn, M.B. (2004), *Migrant resettlement in the Russian Federation: Reconstructing homes and homelands*, Anthem Press, London.

Forsyth, T. (2003), *Critical political ecology: The politics of environmental science*, Routledge, London.

Foster, J.B. (2002), *Ecology against capitalism*, Monthly Review Press: New York.

Garb, P. and Komarova, G. (1999), 'A history of environmental activism in Chelyabinsk', in R.J. Dalton, P. Garb, N.P. Lovrich, et al., *Critical Masses: Citizens, nuclear weapons production, and environmental destruction in the United States and Russia*, The MIT Press, Cambridge (MA), pp. 165-191.

Gare, A. (1996), 'Soviet Environmentalism: The path not taken', in T. Benton (ed.), *The Greening of Marxism*, The Guildford Press, New York/London, pp. 111-128.

Gibson-Graham, J.K. (2004), 'Area studies after poststructuralism', *Environment and Planning A*, Vol. 36, pp. 405-419.

Goldman, M.I. (1972), *The Soils of Progress: Environmental pollution in the Soviet Union*, The MIT Press, Cambridge (MA).

Goskomekologiya (2000), *Gosudarstvennyi doklad "O sostoyanii okruzhayushchei prirodnoi sredy Rossiiskoi Federatsii v 1999 godu"*, Goskomekologiya, Moscow (www.mpr.gov.ru, accessed June 2001).

Goskompriroda (1990), *State of the Environment in the USSR, 1988* (Abridged version of the official report), VINITI, Moscow.

Goskomstat (1996), *Rossiiskii statisticheskii ezhegodnik: ofitsial'noe izdanie*, Goskomstat, Moscow.

Goskomstat (1998), *Okhrana okruzhayushchei sredy v Rossii: ofitsial'noe izdanie*, Goskomstat, Moscow.

Goskomstat (2000), *Promyshlennost' Rossii: ofitsial'noe izdanie*, Goskomstat, Moscow.

Goskomstat (2001a), *Zdravookhranenie v Rossii: ofitsial'noe izdanie*, Goskomstat, Moscow.

Goskomstat (2001b), *Agropromyshlennyi kompleks Rossii: ofitsial'noe izdanie*, Goskomstat, Moscow.

Goskomstat (2002a), *Toplivno-energeticheskii kompleks Rossii: statisticheskii sbornik*, Goskomstat, Moscow.

Goskomstat (2002b), *Chislennost' naseleniya Rossiiskoi Federatsii po gorodam, poselkam gorodskogo tipa i raionam na 1 yanvarya 2002 g.*, Goskomstat, Moscow.

Goskomstat (2003a), *Rossiya v tsifrakh: ofitsial'noe izdanie*, Goskomstat, Moscow.

Goskomstat (2003b), *Osnovnye pokazateli okhrany okruzhayushchei sredy: statisticheskii byulleten'*, Goskomstat, Moscow.

Goskomstat (2004a), *Rossiya v tsifrakh: ofitsial'noe izdanie*, Goskomstat, Moscow.

Goskomstat (2004b), *Osnovnye pokazateli okhrany okruzhayushchei sredy: statisticheskii byulleten'*, Goskomstat, Moscow.

Goskomstat SSSR (1989), *Okhrana okruzhayushchei sredy i ratsional'noe ispol'zovanie prirodnykh resursov v SSSR: statisticheskii sbornik*, "Finansy i statistika, Moscow.

Gossanepidnadzor (1995), 'Natsional'nyi doklad "O sanitarno-epidemiologicheskoi obstanovke v Rossiiskoi Federatsii v 1993 g.', *Zdravookhranenie Rossiiskoi Federatsii*, No. 3, pp. 11-15.

Gossanepidnadzor (2001), *Gosudarstvennyi doklad "O sanitarno-epidemiologicheskoi obstanovke v Rossiiskoi Federatsii v 2000 godu"*, Federal'nyi tsentr gossanepidnadzora Minzdrava Rossii, Moscow.

Gossanepidnadzor (2003), *Gosudarstvennyi doklad "O sanitarno-epidemiologicheskoi obstanovke v Rossiiskoi Federatsii v 2002 godu"*, Federal'nyi tsentr gossanepidnadzora Minzdrava Rossii, Moscow.

Grabener, U. (2001), 'A personal appraisal of the implementation of the Seville strategy in Russia', *Russian Conservation News*, No. 27, pp. 15-17.

Grabher, G. and Stark, D. (eds.) (1997), *Restructuring networks in post-socialism: Legacies, linkages, and localities*, Oxford University Press, Oxford.

Grinevald, J. (1998), 'Introduction: the invisibility of the Vernadskian revolution', in V.I. Vernadsky, *The Biosphere*, Copernicus, New York, pp. 20-32.

Gustafson, T. (1999), *Capitalism Russian-style*, Cambridge University Press: Cambridge.

Hamilton, F.E.I. (1999), 'Transformation and space in central and eastern Europe', *The Geographical Journal*, Vol. 165(2), pp. 135-144.

Hann, C. (ed.) (2002), *Postsocialism: Ideals, ideologies and practices in Eurasia*, Routledge, London.

Hanna J., Parasyuk, N., Makipaa, R. et al. (2000) *Russian Federation: Report on the second national communication of the Russian Federation*, FCCC/IDR.2.RUS, September 27, UNFCCC.

Hanson, P. and Bradshaw, M. (eds.) (2000a), *Regional economic change in Russia*, Edward Elgar, Cheltenham.

Hanson, P. and Bradshaw, M. (2000b), 'Introduction', in P. Hanson and M. Bradshaw (eds.), *Regional economic change in Russia*, Edward Elgar, Cheltenham, pp. 1-16.

Harvey, D. (1996), Justice, nature and the geography of difference, Blackwell, Oxford.

Hayter, R., Barnes, T., and Bradshaw, M.J. (2003), 'Relocating resource peripheries to the core of economic geography's theorizing: rationale and agenda', *Area*, Vol. 35(1), pp. 15-23.

Herrschel, T. and Forsyth, T. (2001), 'Constructing a new understanding of the environment under postsocialism', *Environment and Planning A*, Vol. 33, pp. 573-587.

Hill, M.R. (1999), *Environment and technology in the former USSR: The case of acid rain and power generation*, Edward Elgar, Cheltenham.

Holden, B. (2002), *Democracy and Global Warming*, Continuum, London.

Holloway, D. (1994), *Stalin and the Bomb*, Yale University Press, New Haven.

Holm-Hansen, J. (ed.) (2002), *Environment as an issue in a Russian town*, NIBR-report, No. 11, Norwegian Institute for Urban and Regional Research (NIBR), Oslo.

Hønneland, G. (2003), *Russia and the West: Environmental co-operation and conflict*, Routledge, London.

Hooson, D.J.M. (1962), 'Methodological clashes in Moscow', *Annals of the Association of American Geographers*, Vol. 52, pp. 469-475.

Hooson D.J.M (1968), The Development of Geography in Pre-Soviet Russia *Annals of the Association of American Geographers*, Vol. 58, pp. 250-272.

Humphrey, C. (2002), 'Does the category of 'postsocialist' still make sense?', in C.M. Hann (ed.), *Postsocialism: Ideals, ideologies and practices in Eurasia*, Routledge, London, pp. 12-15.

Humphrey, C. and Mandel, R. (2002), 'The market in everyday life: ethnographies of postsocialism', in R. Mandel and C. Humphrey (eds.), *Markets and moralities: ethnographies of postsocialism*, Berg, Oxford, pp. 1-16.

Ichikawa, N., Tsutsumi, R. and Watanabe, K. (2002), 'Environmental indicators of transition', *European Environment*, Vol. 12, pp. 64-76.

ICPCC (2002), *Tret'e natsional'noe soobshchenie Rossiiskoi Federatsii*, Mezhvedomstvennaya Komissiya Rossiiskoi Federatsii po problemam izmeneniya klimata, Moscow.

IMF, World Bank, OECD and EBRD (1991), *A study of the Soviet economy, Volume 3*, OECD, Paris.

Ioffe, G. and Nefedova, T (2001), 'Russian agriculture and food processing: Vertical cooperation and spatial dynamics', *Europe-Asia Studies*, Vol. 53(3), pp. 389-418.

Ioffe, G. and Nefedova, T. (2004), 'Marginal Farmland in European Russia', *Eurasian Geography and Economics*, Vol. 45(1), pp. 45-59.

Ioffe, G., Nefedova, T., and Zaslavsky, I. (2004), 'From spatial continuity to fragmentation: The case of Russian farming', *Annals of the Association of American Geographers*, Vol. 94(4), pp. 913-943.

IOPRR (2001), 'Sokrashchennyi variant analiticheskogo doklada "Prirodno-resursnyi kompleks Rossiiskoi Federatsii"', *Ispol'zovanie i okhrana prirodnykh resursov v Rossii*, No. 1-2, pp. 3-267.

IOPRR (2002a), 'Statisticheskie dannye o strukture zemel'nogo fonda Rossii', *Ispol'zovanie i okhrana prirodnykh resursov v Rossii*, Nos. 1-2, pp. 83-84.

IOPRR (2002b), 'Ekologicheskaya doktrina Rossiiskoi Federatsii (odobrena rasporyazheniem Pravitel'stva RF ot 31 avgusta 2002 g. No. 1225-r)', *Ispol'zovanie i okhrana prirodnykh resursov v Rossii*, Nos. 7-8, pp. 119-127.

Jancar-Webster, B. (1998), 'Environmental movement and social change in the transition countries', *Environmental Politics*, Vol. 7(1), pp. 69-90.

Josephson, P.R. (1999), *Red Atom: Russia's nuclear power program from Stalin to today*, W.H. Freeman and Company, New York.

Kistanov, V.V. (2000), *Federal'nye okruga Rossii: vazhnyi shag v ukreplenii gosudarstva*, Ekonomika, Moskva.

Kochurov, B.I. (1989), 'Na puti k sozdaniyu ekologicheskoi karty SSSR', *Priroda*, No. 8, pp. 10-17.

Komarov, B. (1978), *Unichtozhenie prirody: obostrenie ekologicheskogo krizisa v SSSR*, Posev, Frankfurt/Main.

Komarov, B. (1980), *The destruction of nature in the Soviet Union*, M.E. Sharpe, Armonk NY.

Konstitutsiya (1997), *Konstitutsiya Rossiiskoi Federatsii*, 'PRIOR', Moscow.

Kornai, J. (1992), *The socialist system: The political economy of communism*, Clarendon Press, Oxford.

Kostenchuk, N.A., Ignatovich, I.V., Galkin, Yu.Yu. and Morev, M.Yu. (1993), *Sostoyanie okruzhayushchei sredy i prirodookhrannaya deyatel'nost' na territorii byvshego SSSR: Ot Stokgol'ma k Rio. Spravochnoe posobie*, Ministerstvo okhrany okruzhayushchei sredy i prirodnykh resursov Rossiiskoi Federatsii, Moscow.

Kotkin, S. (1995), *Magnetic Mountain: Stalinism and civilization*, University of California Press, Berkeley (CA).

Kuus, M. (2004), 'Europe's eastern expansion and the reinscription of otherness in East-Central Europe', *Progress in Human Geography*, Vol. 28(4), pp. 472-489.

Lafferty, W.M. and Meadowcroft, J. (1996) 'Democracy and the environment: congruence and conflict – preliminary reflections', in W.M. Lafferty and J. Meadowcroft (eds.), *Democracy and the environment: Problems and prospects*, Edward Elgar, Cheltenham, pp. 1-17.

Larin. V., Mnatsakanyan, R., Chestin, I. and Shvarts, E. (2003), *Okhrana prirody Rossii: ot Gorbacheva do Putina*, KMK (Scientific Press), Moscow.

Ledeneva, A.V. (1998), *Russia's economy of favours: Blat, networking and informal exchange*, Cambridge University Press, Cambridge.

Lieven, D. (2000), *Empire: The Russian Empire and its rivals*, John Murray, London.

Lisitzin, E.N. (1987), 'The Union of Soviet Socialist Republics', in G. Enyedi, A.J. Gijswijt and B. Rhode (eds.), *Environmental policies in East and West*, Taylor Graham, London, pp. 311-333.

Macnaghten, P. and Urry, J. (1998), *Contested natures*, Sage Publications: London.

Marples, D.R. (2004), 'Chernobyl: A Reassessment', *Eurasian Geography and Economics*, Vol. 45(8), pp. 588-607.

Martin, R. (2001), 'Geography and public policy: the case of the missing agenda, *Progress in Human Geography*, Vol. 25(2), pp. 189-210.

Massey, D. (2005), *For Space*, Sage Publications, London.

Massey Stewart, J. (ed.) (1992), *The Soviet environment: problems, policies and politics*, Cambridge University Press, Cambridge.

Matley, I.M. (1966), 'The Marxist approach to the geographical environment', *Annals of the Association of American Geographers*, Vol. 56, pp. 97-111.

Matley, I.M. (1982), 'Nature and society: the continuing Soviet debate', *Progress in Human Geography*, Vol. 6, pp. 367-396.

Mazurov, Yu.L. (2004), 'Ekologicheskaya politika Rossii v 1990-e gody', *Zelenyi mir*, Nos. 19-20, pp. 15-18.

McCannon, J. (1995), 'To storm the Arctic: Soviet polar exploration and the public visions of nature in the USSR, 1932-1939', *Ecumene*, Vol. 2(1), pp. 15-31.

MChS Rossii, (2004), *Gosudarstvennyi doklad o sostoyanii zashchity naseleniya i territorii Rossiiskoi Federatsii ot chrezvychainykh situatsii prirodnogo i tekhnogennogo kharaktera v 2003 godu*, Emercom, Moscow, (www.mchs.gov.ru/, accessed July 2004).

McIlwaine, C. (1998), 'Civil society and development geography', *Progress in Human Geography*, Vol. 22, No. 3, pp. 415-424.

McNeill, J. (2000), *Something new under the sun: An environmental history of the twentieth century*, Penguin: London.

MEPNR (1994), *State of the Environment of the Russian Federation 1993: National report*, "Image", Moscow.

MEPNR (1995), *Natural Environment in Russia: A brief survey*, MEPNR/Ecos Journal, Moscow.

Micklin, P. (1991), 'The water crisis in Soviet Central Asia', in P. Pryde, *Environmental Management in the Soviet Union*, Cambridge University Press, Cambridge, pp. 213-232.

Mineconomrazvitiya (2002), *Natsional'naya otsenka progressa Rossiiskoi Federatsii pri perekhode k ustoichivomu razvitiyu*, Ministerstvo ekonomicheskogo razvitiya i torgovli Rossiiskoi Federatsii, Moscow (electronic version accessed via the official website of the Ministry for economic development and trade: www.economy.gov.ru/wps/portal, accessed July 2003).

Mineconomrazvitiya (2004), *Osnovnye parametry sotsial'no-ekonomicheskogo razvitiya Rossiiskoi Federatsii na 2005 god i na period do 2007 goda*, Ministerstvo ekonomicheskogo razvitiya i torgovli Rossiiskoi Federatsii, Moscow (electronic version accessed via the official website of the Ministry for economic development and trade: www.economy.gov.ru/wps/portal, July 2004).

MinEkologiya (1992), *Gosudarstvennyi doklad o sostoyanii okruzhayushchei prirodnoi sredy Rossiiskoi Federatsii v 1991 godu*, MinEkologiya, Moscow.

Minpriroda (1996), *Gosudarstvennyi doklad "O sostoyanii okruzhayushchei prirodnoi sredy Rossiiskoi Federatsii v 1995 godu"*, Ministerstvo okhrany okruzhayushchei sredy i prirodnykh resursov Rossiiskoi Federatsii, Moscow.

MinZdrav (1994), 'Gosudarstvennyi doklad o sostoyanii zdorov'ya naseleniya Rossiiskoi Federatsii v 1992 g. (prodolzhenie)', *Zdravookhranenie Rossiiskoi Federatsii*, No. 6, pp. 3-10.

Mirovitskaya, N. (1998), 'The environmental movement in the former Soviet Union', in A. Tickle and I. Welsh (eds.), *Environment and Society in Eastern Europe*, Longman, Harlow (Essex), pp. 30-66.

Missfeldt, F. and Villavicenco, A. (2000), 'The economies in transition as part of the climate regime: recent developments', *Environment and Planning B: Planning and Design*, Vol. 27, pp. 379-392.

Moiseev, N.N. (1999), 'Reflection on the noosphere – humanism in our time', in P.R. Samson and D. Pitt (eds.), *The biosphere and noosphere reader: Global environment, society and change*, Routledge, London, pp. 167-176.

MPR (2001a), *Gosudarstvennyi doklad "O sostoyanii i ob okhrane okruzhayushchei prirodnoi sredy Rossiiskoi Federatsii v 2000 godu"*, Ministerstvo prirodnykh resursov, Moscow.

MPR (2001b), *Adaptirovannyi dlya naseleniya variant Gosudarstvennogo doklada "O sostoyanii okruzhayushchei prirodnoi sredy Rossiiskoi Federatsii v 2000 godu"*, Ministerstvo prirodnykh resursov, Moscow.

MPR (2002), *Gosudarstvennyi doklad "O sostoyanii i ob okhrane okruzhayushchei prirodnoi sredy Rossiiskoi Federatsii v 2001 godu"*, Ministerstvo prirodnykh resursov, Moscow.

MPR (2003), *Gosudarstvennyi doklad o sostoyanii okruzhayushchei prirodnoi sredy Rossiiskoi Federatsii v 2002 godu*, MPR, Moscow.

MPR (2004), *Gosudarstvennyi doklad o sostoyanii okruzhayushchei prirodnoi sredy Rossiiskoi Federatsii v 2003 godu*, MPR, Moscow, (electronic version accessed via the official website of the Ministry for Natural Resources: www.mnr.gov.ru, March 2005).

MPR-LR (2003), *Gosudarstvennyi doklad "O sostoyanii i ispol'zovanii lesnykh resursov Rossiiskoi Federatsii v 2002 godu"*, MPR, Moscow, (electronic version accessed via the official website of the Ministry for Natural Resources: www.mnr.gov.ru, December 2004).

Murphy, J. (2000), 'Editorial: Ecological modernisation', *Geoforum*, Vol. 31, pp. 1-8.

Nee, V. and Stark, D. (eds.) (1989), *Remaking the economic institutions of socialism: China and Eastern Europe*, Stanford University Press, Stanford (CA).

Newell, J. and Wilson, E. (1996), *The Russian Far East: Forests, biodiversity hotspots, and industrial developments*, Friends of the Earth-Japan, Japan.

Newell, J. (2004), *The Russian Far East: A reference guide for conservation and development*, Daniel and Daniel, McKinleyville (CA).

Nilssen, F. and Hønneland, G. (2001), 'Institutional change and the problems of restructuring the Russian fishing industry', *Post-Communist Economies*, Vol. 13 (3), pp. 313-330.

North, R.N. and Shaw, D.J.B. (1995), 'Industrial policy and location', in D.J.B. Shaw (ed.), *The Post-Soviet Republics: A systematic geography*, Longman, Harlow (Essex), pp. 46-65.

Nove, A. (1992), *An economic history of the USSR: 1917-1991* (Final edition), Penguin, London.

O'Hara, S.L. (2004), 'Great game or grubby game? The struggle for control of the Caspian Sea, *Geopolitics*, Vol. 9(1), pp. 138-160.

OECD (1996), *Environmental Information Systems in the Russian Federation: An OECD assessment*, OECD, Paris.

OECD (1999), *Environmental performance reviews: Russian Federation*, OECD, Paris.

OECD (2002a), *OECD Environmental Compendium 2002: Air*, OECD, Paris.

OECD (2002b), *OECD Environmental Compendium 2002: Waste*, OECD, Paris.

OECD (2002c), *OECD Environmental Compendium 2002: Agriculture*, OECD, Paris.

Oldfield, J. (1999), 'Socio-economic change and the environment - Moscow city case study', *The Geographical Journal*, Vol. 165(2), pp. 222-231.

Oldfield, J.D. (2000), 'Structural economic change and the natural environment in the Russian Federation', *Post-Communist Economies*, Vol. 12(1), pp. 77-90.

Oldfield, J.D. (2001), 'Russia, systemic transformation and the concept of sustainable development', *Environmental Politics*, Vol. 10(3), pp. 94-110.

Oldfield, J.D., Kouzmina, A. and Shaw, D.J.B. (2003), 'Russia's involvement in the international environmental process: A research report', *Eurasian Geography and Economics*, Vol. 44(2), pp. 157-168.

Oldfield, J.D. and Shaw, D.J.B. (2002), 'Revisiting Sustainable Development: Russian cultural and scientific traditions and the concept of Sustainable Development', *Area*, Vol. 34(4), pp. 391-400.

Oldfield, J.D. and Shaw. D.J.B. (in press), 'V.I. Vernadsky and the noosphere concept: Russian understandings of society-nature interaction', *Geoforum.*

Oldfield, J.D. and Shaw, D.J.B. (forthcoming) 'Russian understandings of society-nature interaction: Historical underpinnings of the sustainable development concept', in Arja Rosenholm and Sari Autio-Sarasmo (eds.), *Understanding Russian nature: Representations, values and concepts*, Kikimora Publications (Series B), Helsinki.

Oldfield, J.D. and Tickle, A. (2005), 'Foreign economic relations and the environment: a historical approach', in D. Turnock (ed.), *Foreign Direct Investment and Regional Development and the Former Soviet Union (The Memorial volume for Frank Carter)*, Ashgate Press, pp. 123-139.

Paehlke, R. (1996), 'Environmental challenges to democratic practice', in W.M. Lafferty and J. Meadowcroft (eds.), *Democracy and the environment: Problems and prospects*, Edward Elgar: Cheltenham, pp. 18-38.

Pallot, J. and Shaw, D.J.B. (1981), *Planning in the Soviet Union*, Croom Helm: London.

Pallot, J. and Nefedova, T. (2003), 'Geographical differentiation in household plot production in rural Russia', *Eurasian Geography and Economics*, Vol. 44(1), pp. 40-64.

Pavlinek, P. and Pickles, J. (2000), *Environmental Transitions: Transformation and ecological defence in central and eastern Europe*, Routledge, London.

Peterson, D.J. (1993), *Troubled Lands: The legacy of Soviet environmental destruction*, Westview Press, Boulder (CO).

Peterson, D.J. (1995a), 'Russia's environment and natural resources in light of economic regionalization', *Post-Soviet Geography*, Vol. 36(5), pp. 291-309.

Peterson, D.J. (1995b), 'Building bureaucratic capacity in Russia: Federal and regional responses to the post-Soviet environmental challenge', in J. DeBardeleben and J. Hannigan (eds.), *Environmental security and quality after communism: Eastern Europe and the Soviet successor states*, Westview Press, Boulder (CO), pp. 107-126.

Peterson, D.J. and Bielke, E.K. (2002), 'Russia's Industrial Infrastructure: A Risk Assessment', *Post-Soviet Geography and Economics*, Vol. 43(1), pp. 13-25.

Pickvance, C. (2003), *Local environmental regulation in post-socialism: A Hungarian case study*, Ashgate, Aldershot.

Porritt, J. (1984), *Seeing Green: the politics of ecology explained*, Basil Blackwell, Oxford.

Programma (2001), *Programma sotsial'no-ekonomicheskogo razvitiya Rossiiskoi Federatsii na srednesrochnuyu perspektivu (2002-2004 gody)*, Approved by ruling of the Russian government, 10/07/2001, No. 910-r, Moscow.

Pryde, P.R. (1972), *Conservation in the Soviet Union*, Cambridge University Press, Cambridge.

Pryde, P.R. (1991), *Environmental Management in the Soviet Union*, CUP, Cambridge.

Pryde, P.R. (1994), 'Observations on the mapping of critical environmental zones in the former Soviet Union', *Post-Soviet Geography*, Vol. 35(1), pp. 38-49.

Rodgers, A. (ed.) (1990), *The Soviet Far East: Geographical perspectives on development*, Routledge: London.

Rotfel'd, I.S. and Medvedovskii, S.Ya. (2003), 'Toplivno-energeticheskie resursy Rossii: retrospectiva i perspektiva', *Ispol'zovanie i okhrana prirodnykh resursov v Rossii*, Nos. 11-12, pp. 31-47.

Rozenberg, G.S., Gelashvili, D.B., Krasnoshchekov, G.P. (1996) 'Krutye stupeni perekhoda k ustoichivomy razvitiyu', *Vestnik Rossiiskoi Akademii Nauk*, Vol. 66(5), pp. 436-440.

Saiko, T. (2001), *Environmental crises: Geographical case-studies in post-socialist Eurasia*, Prentice Hall, Harlow (Essex).

Sachs, W. (1999), Planet dialectics: Explorations in environment and development, Zed Books, London.

Shaw, D.J.B. (1985), 'Spatial dimensions in Soviet central planning', *Transactions of the Institute of British Geographers – NS*, Vol. 10, pp. 401-412.

Shaw, D.J.B. and Oldfield, J. (1998), 'The natural environment of the CIS in the post-Soviet period', *Post-Soviet Geography and Economics*, Vol. 39(3), pp. 164-177.

Shaw, D.J.B. (1999), *Russia in the modern world: a new geography*, Blackwell, Oxford.

Singleton, F. (ed.) (1976), *Environmental misuse in the Soviet Union*, Praeger Publishers, New York.

Smith, A. (2002), 'Imagining geographies of the 'new Europe': geo-economic power and the new European architecture of integration', *Political Geography*, Vol. 21, pp. 647-670.

Smith, A. (2004), 'Regions, spaces of economic practice and diverse economies in the 'new Europe', *European Urban and Regional Studies*, Vol. 11(1), pp. 9-25.

Smith, A. and Swain, A. (1998), 'Regulating and institutionalising capitalisms: The micro-foundations of transformation in Eastern and Central Europe', in J. Pickles and A. Smith (eds.), *Theorising transition: The political economy of post-communist transformations*, Routledge, London, pp. 25-53.

Smith, G. (1999), *The Post-Soviet states: Mapping the politics of transition*, Arnold, London.

Smol'yakova, T. (1997), 'Kommentarii - postanovlenie pravitel'stva Rossiiskoi Federatsii ob informatsionnykh uslugakh v oblasti gidrometeorologii i monitoringa zagryazneniya okruzhayushchei prirodnoi sredy', *Rossiiskaya Gazeta*, 2 December, p. 4.

Soos, K.A., Ivleva, E. and Levina, I. (2002), 'Russian Manufacturing Industry in the Mirror of its Export to the European Union', *Russian Economic Trends*, Vol. 11(2), pp. 31-43.

Staddon, C. (2001), 'Restructuring the Bulgarian wood-processing sector: linkages between resource exploitation, capital accumulation, and redevelopment in a postcommunist locality', *Environment and Planning A*, Vol. 33, pp. 607-628.

Stark, D. and Bruszt, L. (eds.) (1998), *Postsocialist pathways: Transforming politics and property in east central Europe*, Cambridge University Press, Cambridge.

Stepanitsky, V. (2001), 'The particulars of Russian biosphere reserves', *Russian Conservation News*, No. 27, pp. 13-14.

Swain, A. (2002), 'Broken networks and tabula rasa? Lean production, employment and the privatisation of the East German automobile industry', in A. Rainnie, A. Smith, and A. Swain (eds.), *Work, Employment and Transition: Restructuring livelihoods in post-communism*, Routledge, London, pp. 74-96.

Taga, L.S. (1976), 'Externalities in a command system', in F. Singleton (ed.), *Environmental misuse in the Soviet Union*, Praeger Publishers, New York, pp. 75-100.

Tickle, A. and Welsh, I. (eds.) (1998), *Environment and society in eastern Europe*, Longman, Harlow (Essex).

Tickle, A. (2000), 'Regulating environmental space in socialist and post-socialist systems: nature and landscape conservation in the Czech Republic', *Journal of European Area Studies*, Vol. 8(1), pp. 57-78.

Ukaz (1997a), 'Polozhenie o ministerstve prirodnykh resursov Rossiiskoi Federatsii', *Rossiiskaya gazeta*, June 3, p. 4.

Ukaz (1997b), 'Polozhenie o gosudarstvennom komitete Rossiiskoi Federatsii po okhrane okruzhayushchei sredy', *Rossiiskaya gazeta*, June 10, p. 5.

Ukaz (2004), 'Ukaz Prezidenta Rossiiskoi Federatsii ot 9 Marta 2004 g. N 314 O sisteme i strukture federal'nykh organov ispolnitel'noi vlasti', *Rossiiskaya*

Gazeta, March 11 (www.rg.ru/2004/03/11/federel-dok.html, accessed July 2004).

UNEP (2002), *Global Environment Outlook 3: Past, present and future perspectives*, UNEP/Earthscan, London.

UNFCCC (2000), *Russian Federation: Report on the in-depth review of the second national communication of the Russian Federation*, September 27, FCCC/IDR.2/RUS (www.unfccc.int/, accessed January 2004).

Ustoichivoe razvitie (1996), Rossiya na puti k ustoichivomu razvityu, REFIA, Moscow.

Unwin, T., Pallot, J. and Johnson, S. (2004) 'Rural change and agriculture', in M. Bradshaw and A. Stenning (eds.), *East central Europe and the former Soviet Union*, Pearson Education, Harlow (Essex), pp. 109-136.

Van Buren, L. (1995), 'Citizen participation and the environment in Russia', in J. DeBardeleben, and J. Hannigan (eds.) *Environmental security and quality after communism: Eastern Europe and the successor states*, Westview Press, Boulder (CO), pp. 127-137.

Vasil'eva, M.I. (2002), *Novoe v Federal'nom Zakone "Ob okhrane okruzhayushchei sredy". Kommentarii*, REFIA, Moscow.

Verdery, K. (2002), 'Whither postsocialism', in C. M. Hann (ed.), *Postsocialism: Ideals, Ideologies and Practices in Eurasia*, Routledge, London, pp. 15-22.

Vernadsky, W.I. (1945), 'The biosphere and the noösphere', *American Scientist*, Vol. 33(1), pp. 1-12.

Vernadsky, V.I. (1998), *The Biosphere*. New York, Copernicus.

Vucinich, A. (1970), *Science in Russian culture, 1861-1917*, Stanford University Press, Stanford (CA).

WCED (World Commission on Environment and Development) (1987), *Our Common Future*, Oxford University Press, Oxford/New York.

Wegren, S. (2002), 'Russian agrarian policy under Putin', *Post-Soviet Geography and Economics*, Vol. 43(1), pp. 26-40.

Weiner, D.R. (1988) *Models of nature: Ecology, conservation and cultural revolution in Soviet Russia*, University of Pittsburgh Press, Pittsburgh (Pa).

Weiner, D.R. (1999), *A Little Corner of Freedom: Russian nature protection from Stalin to Gorbachëv*, University of California Press, Berkeley (CA).

Willis, K. (2004), '"Distant Geographies", international understanding and global co-operation', *Geoforum*, Vol. 35, pp. 399-400.

Wilson, E. (2000), *North-eastern Sakhalin: local communities and the oil industry*, Russian Regional Research Working Paper No. 21, The University of Birmingham and University of Leicester.

Wilson, E. (2002), 'Est' zakon, est' i svoi zakony: Legal and moral entitlements to the fish resources of Nyski Bay, North-Eastern Sakhalin', in E. Kasten (ed.), *People and the land: Pathways to reform in post-Soviet Siberia*, Dietrich Reimer Verlag, Berlin, pp. 149-168.

Wilson, E. (2003), 'Freedom and loss in a human landscape: multinational oil exploitation and the survival of reindeer herding in north-eastern Sakhalin, the Russian Far East', *Sibirica*, 2003, Vol.3(1), pp. 21-47.

Wolfson, Z. (1991), 'Foreword', in Pryde, P.R., *Environmental management in the Soviet Union*, Cambridge University Press, Cambridge, pp. xv-xvi

World Bank (1997), *Russia: Forest policy during transition*, World Bank, Washington D.C.

World Bank (2002), *Transition: The First Ten Years. Analysis and lessons for eastern Europe and the former Soviet Union*, World Bank, Washington D.C.

World Bank (2003), *World Development Report 2003. Sustainable Development in a Dynamic World: Transforming Institutions, Growth, and Quality of Life*, The World Bank/Oxford University Press, New York.

Yanitsky, O. (1993), *Russian Environmentalism: Leading Figures, Facts, Opinions*, Mezhdunarodnyje Otnoshenija Publishing House, Moscow.

Yanitsky, O.N. (2000), *Russian Greens in a risk society: A structural analysis*, Kikimora Publications (Series B), Helsinki.

Yanitsky, O. (2002), *Rossiya: ekologicheskii vyzov (obshchestvennoe dvizhenie, nauka, politika)*, Sibirskii khronograf, Novosibirsk.

Zamparutti, T. and Gillespie, B. (2000), 'Environment in the transition towards market economies: an overview of trends in Central and Eastern Europe and the New Independent States of the former Soviet Union', *Environment and Planning B: Planning and Design*, Vol. 27, pp. 331-347.

Ziegler, C.E. (1992), 'Political participation, nationalism and environmental politics in the USSR', in J. Massey Stewart (ed.), *The Soviet environment: problems, policies and politics*, Cambridge University Press, Cambridge, pp, 24-39.

Index